Machines Von Aryans

By Sri Kutupananda Nath

Author Contact
vamapathofshakti@gmail.com

To
Scientiests & Engineers of Ancient India whose names & works have gone into oblivion

Contents

Introduction

If we truely say the one field where Germans under leadership of NAZI party advanced then it that is the field of hardcore Mechanical engineering . From manufacturing innovative extra resistant and hard materials to designing jet engines for fighter planes , from luxurious cars of Mercedez to heavy machinery of ThysuunKrup , Bosch German manufcaturing set a standard which is still considered top notch in hardcore industry of machining , automobile & process industries .

Under the support & encouragement from leaders like Himmler himself private corpos in Deustchland were able to take leaps in inventions in engineering . If we take a look at hidden or abandoned history of German weapons , this leap was also in the field of Unified Field theory & Anti Gravity propulsion technology . Leading scientists in Germany like Schauberger , Schumann who were able to invent anti gravity propulsion . With mechanical application of Unified field theory , Germans arrived at possibility of Teleportation & Time travel in parallel worlds . Though in our space time Hitler had lost the war & now Jewish sciences of Relativity & String Theory rules everywhere , time before the 1945 April , there was considerable progress in connecting quantum theory with unified field theory to practically achieve time space travel without imaginary worm hole or black hole fantasies which run rabid in currunt Astrophysicists . We can say in simple words , the day we shifted from Mechanical engineering to Jewish sciences of theorotical physics , we are stuck at wall of unending mathematics .

Not only we were forced against this black wall of theoretical physics , we are put under deception by continuous growing web of electronics . Post WWII , electronics is encroaching our lives like termites . It started from discovery of transister by Von Neumann (who came from other timeline allegedly after Philadelphia Experiment) and then it expanded with that numerological law of Moore on size of transister . Occutists Philosopher Rudolf Steiner called electronics as advancement of Ahirman , the demon of cold fire . Symbolism of cold fire demon is for cold temp movement of electric currunt in digital circuits in relation to hot movement in analog electrical devices . This Devil of neitherworld of forever cold has entrapped our lives now . Manufacturing , transport, entertainment , even in simpler functions like a keeping a note, sending a letter has became digitalized over the years . Now we see no escape from this. But was this destined to happen ?

From Cathode tube of color TV , was it a destiny that we must had progress to transiter based technology ? Was it forced on us ? US company Philips Co which was first to use transister for color TV , got this new technology from suspicious sources . Rise of this digital tech is post WWII and specially post Rosewell UFO crash of July 1947 . Now when data runs in high speeds of photons in undersea optical cables and when we are at door of personalized quantum computer , looking back at color TV of 1947 will sound crazy .

We are at year 2020 and transister size has reached its last limit of 7 nm . From here its leap into unknown of quantum computing . Will it affect our world in matters of time space ? Will it alter our reality ? Effects like Mandella effect are outcomes of this ? When data downloads at out phone in sizes of Bytes , specially in Torrent downloads where you can observe electrons going in & out into your system , you are looking at electrons under rule of Heisenberg's uncertainity principle . That electron is there because your are viewer of it at that space-time . You are too close of quantum leap into parallel worlds , at a least there are eight parallel worlds which are mingling when data downlads . You might have noticed , download speed reduces when you are watching the speed in torrent . This is the most Personalized quantum effect , nothing else . Smartphones in our hands are potential Time Space Traveling devices which are altering our reality in such a slow manner that change is unnoticable for most people . But if we look closely enough then phenomen like Mandela effect comes into the picture . Technology is magic in disguise . When Hitler lost the war to Ahirbudhnya of Patala or Ahirman of colder hells , we got under this magic of electronics .

But things were different before fall of Hitler or are may be still different in parallel world where Hitler won the war . There are few stories of this possibility in fiction & in non mainstream conspiracy circles . Advance of analog mechanical technology without indulgance in transisters is to be found in two places . One in such parallel world where Hitler had won the war and Second in advanced Vedic technology of Indians . There is obvious reason also why in these two places . Reason is , from India only this technology was taken to Germany in early 1920s . Anti gravity propulsion of UFOs, figher planes , electric vehicles , railways, telephone/telegram to time space travel machines under Himmler's Annerbele , it was taken from secret Indian sources in early 20th century . One such source was Subbaraya Shastri of Bangalore . Stories say he got these ancient technological books in trance & he merely dictated it . But that can be a cover up to hide from where the truly these books came . G R Josyer who published few of Subbaraya Shastri's works in 1970s was very close to one secret occult society in south india called Suddha Dharma Mandala . Infact he translated few of works of this society . One such work was Pranavavada . Pravanavada which is a secret sanskrit philosophical book not found in any extant catalogues of manuscripts of sanskrit books in India or Europe , was a unique finding . This and few other philosophical books were dictated to Parameshvari Das, member of Theosophical Society by one mysterious blind man named Pandit Dhanraj .

Pandit Dhanraj was a blind young man in early thirtees when he met Parameshvari Das of Theosophical Society in Varanasi . He dictated many texts from his memory . Later Suddha Dharma Mandal from Chennai, Tamilnadu published these books . Source of Subbaraya Shastri is some unknown Sannyasi Guru who gave Shastriji power to get ancient lost books in trance and dictate them . Shastri was working in Chennai with one astrological magazine editor to publish his works . Said editor was also in Chennai & G R Josyer who after decades published books dictated by Shastriji . Lets come to Pandit Dhanraj . He gave his native place as village in Uttara Badari in Himalayas where he said many Brahmin families exist which had real texts on ancient philosophy and sciences . As per him , texts published as great works of sanskrit likes Vedas or Puraanas or philosohical texts like Adi Sankaracharya's commentaries were not genuine works but written by later authors . He gave lists of such original texts . In one such list which he gave to District Magistrate in Varanasi after few years have many scientific works in Sanskrit which are yet unknown . Though names of works given by Pandit Dhanraj do not contain works given by Subbaraya Shastri , considering modus operandi & secret nature of books and most importantly common factor who is publisher G R Josyer , there is greater possibility that works like Brihad Vimana Shastra , Amshubodhini were given by Pandit Dhanraj only . Other than these two works & half part of Kritaka Vajra Nirmana , from Shastriji we have only excerpts of other works of which only names are mentioned . Few things are also noteworthy . Suddha Dharma Mandala was trying to bring an Avatar in 1920s who was about to change the future of humanity . Headquarters of Suddha Dharma Mandala was in Uttara Badari from where Pandit Dhanraj came and that place is the area where mythical village of Kalapa belongs . Kalapa is said to be the secret village where future avatar Kalki is being trained . Students of NAZI occultism or Hitler Occultism know that ardent devotte of Hitler, Savitri Devi believed that Hitler was Kalki avatar . Savitri Devi was german & she wrote many books on Hitler . There was another orginzation under Theosophical Society , Order of the Star , lead by Annie Besant which was promoting ideas of avatar of Maitreya in same time . And there were Sri Aurobindo & Mirra Alfassa who were trying to bring Superhuman Avatar in 1926 . Mirra Alfassa , Jew occultist who became very close to Sri Aurobindo in 1910s . Sri Aurobindo was Indian freedom fighter from Bengal who had close connections with high ranks in German army before & after WWI . Role of Sri Aurobindo in all of this will be explained later . Here we should know that there were multiple orgs waiting for avatar like Kalki or Maitreya . Is Kalki the enemy of Maitreya ? Maitreya is more related to 1st avatar of Vishnu called Matsya . As per writer of this book , who is occultist , Matsya-Maitreya avatar comes after killing of Kalki Avatar . Kalki is the warrior Avatar who takes the world to the doom . He is worst than 8th avatar Krishna in terms of deception & destruction . So were two opposite forces trying to bring these two Avtars against each other ? Was Suddha Dharma Mandala behind Avatar in Hitler & Was the same org gave ancient technology to Hitler after he came to power in Germany ? Or Was Sri Aurobindo behind manifestaion of Hitler Avatar ? Did he gave technology to Hitler ? Writer of this book has found by occult means connection between Sri Aurobindo & Subbaraya Shastri . Sri Aurobindo got works on ancient technology from Subbaraya Shastri when Shastri was Professor in Mysore University . What was the role of Jewish Mirra Alfassa in this ? I should note here that what present writer found is opposite to the official version given by followers of Sri Aurobindo & Sri Aurobindo Mirra themselves about their stand in WWII . They openly supported the British in war against Hitler . We have no official versions of pre WWII . Aftet that most of their followers were from US-UK-France or countries who were enemies of Hitler . Their famous disciple Satprem who wrote huge 13 volumes of Mother's Agenda was alleged victim of a concentration camp . Obviously what he or other wrote will be against Hitler &Germany .

Because the secrecy of operations we will not have hard evidence of this technology transfer from India to Germany in early 19th centrury . But we have laws of synchronocity on our sides . And they indicate that someone from south india , very close to Pondicherry-Chennai region were operating & observing the Rise & fall of third Reich. His goal was to bring Supramental new race of next species on the earth which will be next step in the evolution . This race was supposed to come after the victory of Adolf Hitler . Hitler himself in his autobiography mentions about coming of new race . As Hitler lost the war , this new Supramental race didnot come. Seer from Pondicherry left his body for never to come back , Subhashchandra Bose, INS chief lost in fog of unknown and Hitler, Himmler, Gobels had to leave the bodies by suicides. Post Nuremberg truth is different than reality of pre 1945 April . Along with Jewish fake sciences of therotical physics & astrophysics , we have to bear fake history of jew genocides . Under this garb, almost all evidence of India's role in rise of Hitler is gone . Infact that technology itself is gone or perhaps lies in secret files of US Pentagon which they took from Germany. Only remains few leaves here & there . Such few leaves we are presenting here .

Machines of Aryans is the booklet introducing the new readers of this subject to ancient machines of Indians which were based on non-electronics , pure mechanical-electrical technology . Sixty three such machines / vehicles are presented here first time along with their sanskrit names , description & 3D drawings . This document was found in extant manuscripts of Subbaraya Shastri who as mentioned above gave us Vimana Shastra & Amshubodhini . This document is short one but gives many hints to technological advances of Indians. From Rail locomotive to disc shaped UFO, from drilling , mining machines to Radars, from 7D entertainment devices to 25km illuminating artifical Sun , from electric bikes , cars to spy & destructive army vehicles like tank , student of ancient technology will find here the treasure . Sanskrit names of machines are kept as it is & english description is also kept as it is as given by Shastriji. Units of speeds & weight are different than currunt . Its conversion is given .

 We thank all writers & publishers of ancient vedic technology and NAZI Germany history .This is very small attempt to keep the records alive . Though this part of history sounds magical , its more true than mainstream history in published history books on WWII . This small booklet will become a portal for many to come to other side .

Sri Kutupananda Nath
Pondicherry

Units of Measurement

1 Krosha	2 miles
1 Dhanu	3 to 4 m
1 Hasta	400 to 500 mm
1 rattal	2.4 Kg
1 Yojana	1 miles

Chapter I

Machines of Aryans

1. Panchamukha Yantra पंचमुख यंत्र

Panchamukha Yantra is a car with doors to east, to the southwest, to north and on the top. It weighs about 170 Ratals. It transports about thousand Ratals. With the aid of the electricity it can travel at five Kroshas per hour. It is used as a way of transport for men and for wars. Because of the carts is conducted by a spirit called Gaja it is called Gajaakarshanna Panchamukha Ratha.

2 Mrugaakasrshana Yantra मृगाकर्षण यंत्र

This yantra is designed in the shape of animals as oxes, donkeys, horses, camels, elephants and so on.

3 Chaturmukka Ratha Yantra चतुर्मुख रथयंत्र

this cart has some openings on four sides. It weighs 120 Ratals. It can be conducted with every type of oil (fuel, ndr) possibly what is coming from coconut shells, or through using the electricity. आयटी travels at six Kroshas per h It can be conducted with every type of oil (fuel, ndr) possibly what is coming from coconut shells, or through using the electricity. It travels at six Kroshas per hours. it is used to travel, during wars and to transport some objects.

4 Trimukha Ratha Yantra त्रिमुख रथ यंत्र

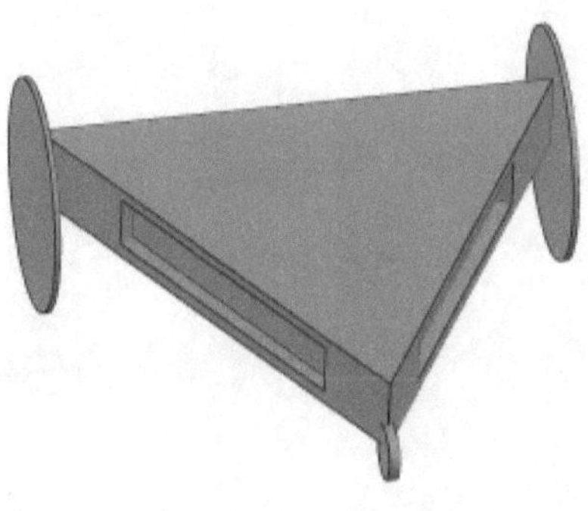

this cart weighs 116 Ratals. It has doors, at the bottom, at the top and on one side. it can support a weight of 600 Ratals. It is conducted by the aid of an oil extract from the root knotted of the Simha-Krantha and from that one extracted on the stems of a type of grass. If these oils are not available, we can use in their place the electricity. It is used for the same purposes of the above mentioned cart Cahkra mukha Ratha Yantra.

5 Dwimukha Yantra द्विमुख यंत्र

it weighs 80 Ratals. It has some openings on east and west. It is governed by a wheel mounted with screws. it travels at 3 Kashas per hours. It can carry a weight of 300 rates. It is used for the purposes referred to above.

6 Ekamukha Ratha Yantra एकमुख रथ यंत्र

 this cart has only one door. It weighs 48 Ratals. It has a weigh of 200 Ratals. It travels thanks to the aid of the oils extracted from the seeds of Kancha Thoola and Sovlaalika or through the electricity. The speed is 1 Kosha per hours. It is used for the above mentioned purposes. This is one of the tank model

7 Simhaasya Ratha Yantra सिंहास्य रथ यंत्र

 this cart has the anterior side with an appearance of a lion. It owns two doors. It weighs 75 Ratals. it carries a weigh of 50 Ratals. It can travel both on the ground that on the air. It possesses the ability to expand and to contract. It is used for the purposes mentioned above.

8 Vyaaghraasya Ratha Yantra व्याघ्रास्य रथ यंत्र

 it is modeled in the shape of a tiger, it owns wings, weighs 64 Ratals, carries 200 Ratals of weight. It travels in the air expanding his wings with the electric power, but contracts its wings with the power of the steam. It is used for the purposes mentioned above.

9 Dolamukha Yantra दोलामुख रथ यंत्र

 it is modeled in the shape of a litter, has two doors, weighs 50 Ratals. [t travels to 3 Kroshas per hour. It is governed by the aid of the electricity and by an oil, namely from the Shilyusha extracted from the wine.

10 Kurmamukha Ratha Yantra कूर्ममुख रथ यंत्र

 is modeled in the shape of a turtle. It owns two small doors, weighs 32 Ratals. it is used only to spy.

11 Ayah Prasaarana Yantra अय: प्रसारण यंत्र

 It moves by the electricity this one particularly travels on the iron lines spread on the Ground. it can be constructed to contain from the 40 to 80 wheels. It seems something like a train, and weighs 4,000 Ratals. It can bring 25,000 Ratals. It travels at 3 Kroshas per hour with the power of the electricity. It is used to transport men and goods from a place to another.

12 Panchamukee Yantra पंचमुखी यंत्र   this can has five sides. It weighs 115 Ratals, can carry 12,000 of them. It has another can that allows for the five doors to open and to close by themselves. It is governed by the electricity. It travels at 4 Kroshas per hours. It is used for the purposes mentioned

13 Eka Chakra Yantra एकचक्र यंत्र

 It carries only a wheel. It is shaped as a trap. it weighs 103 Ratals The movment is given by the gears. It travels at 3 Kroshas per hour.

14 Trimukhi Yantram त्रिमुखी यंत्र

 thIs machine has three sides It contains three compartments that can be separated. It weighs 1.000 Ratais. It travels on the water. The three companments are arranged in such a way that they can travel with the second compartment when first is damaged and if it is also the second. the third can safeguard the contents separating itself from the other if it becomes necessary . It can travel in air and water

15 Jrumbhala Yantra जृंभल यंत्र

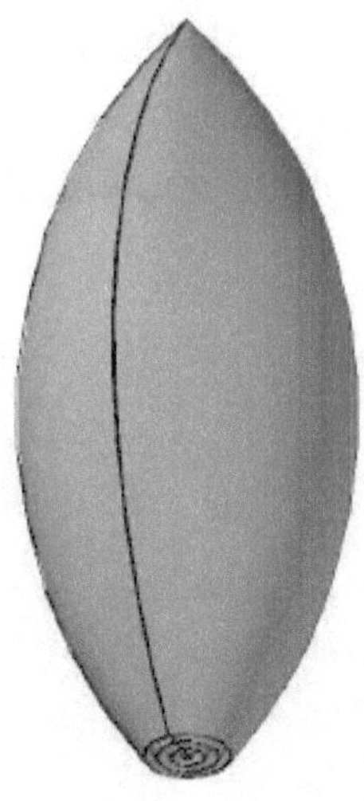

this machine can has an opening at the bottom. It is shaped as a closed umbrella. It is made oi a thick and waterproof fabric produced from the juice of five trees or Pachavarga Ksheera Vriksha. It weighs 42 Ratals. It carries 300 ratals. It can expand in the shape of a pavillion through a screw impeller housed inside. So it can conttact in its original form through another screw. It looks like a flag It is used for secret surveyilance . It can travel to 6 kroshas as per hour thanks to the electric power

16 Goodha Gamam Yantra or Disc UFO गूढ गमन यंत्र

this can accommodate only three people. Its weighs is half oi a Mound. It looks him an orlinary tower. It conlains five keys . It can travel on the ground and in the air. Its motion is almost invisible. it can travel at 8 Kroshas per hour. It operates on the power of a fuel oil called SinjurilIa . It Is used for secret travels.

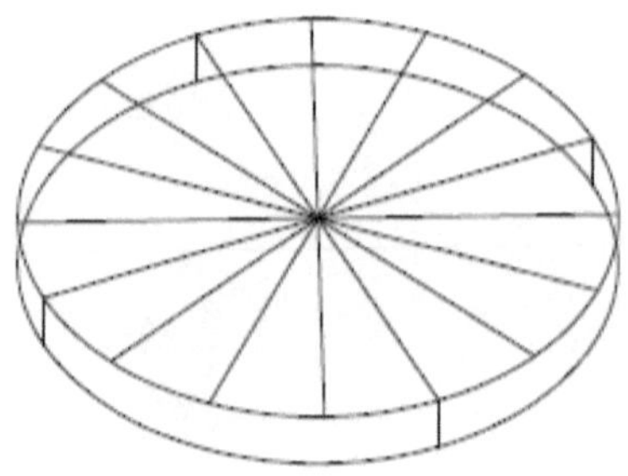

17 Vryajika Yantra व्याजिका यंत्र

This cart is made of glasses of mica. It has sixteen openings and weighs 3 Ratals . . It appears as a shimng light no one can understand that it is a cart. omeone has to go near it. the sparkling light can kill him . it tan travel oth on water and land . It works on electric power of sunlight , it can ravel at 12 Kroshas per hour. It is used travels , wars and money transport

8 Indranee Yamra इंद्राणी यंत्र

this can is constructed hard paper created from grass that belongs to the Maunjmama: the 3'. 9",! '2'. 30" e 42" two of grass named of Pishangalnlgamunja. Pingalamunja. Rajjumuniand so on. This can cannot e destroyed by fire or water. It is extremely light and strong. It cantravel 5 kroshas per hour with the aid of wheels moved by the wind. It brings 00 Ratals.

19 Vishwaavasu Yantra विश्वावसू यंत्र

his cart owns two doors. It weighs 148 Ratals: it brings 3,000 Ratals. Thanks to the help of the steam can travel to 2 Kroshas and half per hour. t can go back and forth. It can be expanded or contracted. It contains even keys. It is used for the purpose mentioned above.

20 Sourambhaka Yantra सौरंभक यंत्र

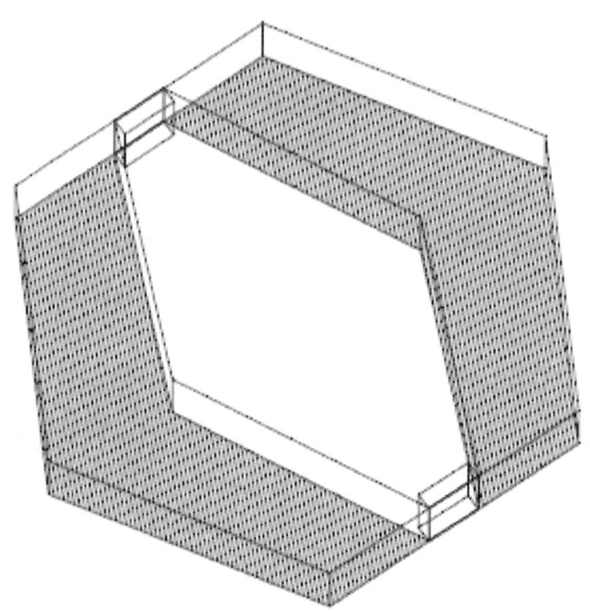

this can owns three floors. There are secret seats for 400 people in each of the three levels. The seats are not normally visible. The levels can only be perceived. it weighs 230 Ratals. it brings 36,000 Ratals. It can travel thanks to the aid of the electricity or of the steam, or with the aid of the spirits of the seventh type of wine. It can travel to 32 Kroshas per hour. It is useful to transport some men or some objects in war.

21 Sphotanee Yantram स्फूटणी यंत्र

this can owns only one door, it weighs 50 Ratals. It brings 200 Ratals. It sails in the water, as a bubble water, sometimes it can rise over the water and sometime can dive under the water. It moves itself with the power of the steam or through the spirits of the Kanajala Kshaara. It travels at 4 Kroshas per hour. It is used for the marine espionage.

22 Kasava Yantra कासव यंत्र this can is modeled as a turtle. It weighs 500 Ratals. It brings 8,000 Ratals. lt owns two doors. It travels under the water. It is used for the purposes mentioned above.

23 Kamala Yantra कमला यंत्र

 this can is modeled as a lotus. It owns four doors. It weighs 69 Ratals. It brings 800 Ratals. A pole is fixed in its central point and encloses some keys that let the can expand and to contract as the lotus does. Trough the steam or the electricity it can travel to 24 Kroshas per houts. It used to travel to remote islands.

24 Taraamukha Yantra तारामुख यंत्र

this cart contains a face of seven keys that are shining as a steel. It has twelve doors, it weighs 2,000 Ratals. It brings 25,000 Ratals. If it is activated the first among the seven keys, a melodious music is accompanied with any type of musical instruments, will be heard by all the people that are in its inside: if it is activated the second one it will be possible to see a dramatic view or a scenario of action: if you active the third, a gentle stream of water will flow through the occupants, so that they can make use of the liquid as they wish: activating the fourth, some tables with flowers, perfumes, camphor, bananas etc. will be available for all the occupants so that they can worship the God: activating the fifth, some trays with excellent food will be offered to them and while they are eating, activating the sixth, the dishes will turn through some wires: activating the seventh, some beds will be available for everyone. If the keys will be repositioned as they were before, everything will vanish. It can travel with the aid of the steam or of the electricity at 4 Kroshas per hour. It is used for the purpose mentioned above.

25 Rohinee Yantra रोहिणी यंत्र

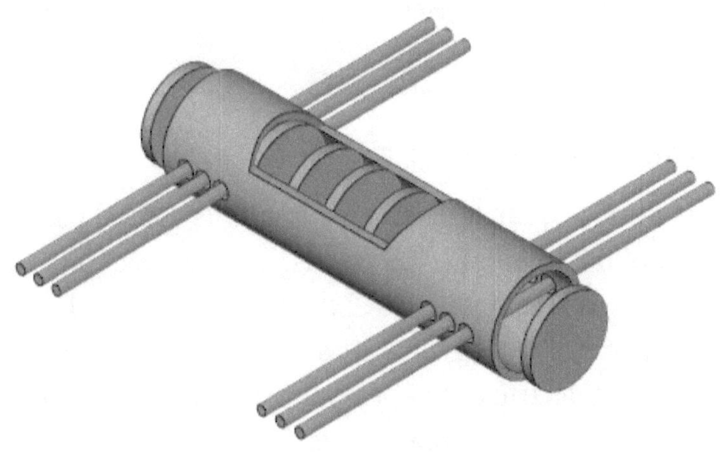

this can is modeled on the shape of a hollow bamboo and it owns the same color. It weighs 3,000 Ratals. It brings 50,000 Ratals. It contains 500 compartments where the explosive powder, the projectiles, the weapons etc. can be preserved. However through some external firebreaks nothing may be burned or damaged, as the fire will be suppressed by the nature of the metal with which is composed. With the help of the steam or of the electricity it may run at 6 Kroshas per hour. It is used especially during the wars.

26 Raakaasya Yantram राकास्य यंत्र

A glorious light as the light of the moon will come out of the cart, every three hours. This light will light up till a distance of 64 Kroshas within which everything that will be inside its beam of light will be clearly visible. It weighs 1 Ratal. Inside the can there is a wheel that continuously runs on the right as the sun. It can travel on the land, on the water and on the air. It is useful to discover some objects in the distance. With the help of the spirits of the wine of the sixteenth type it can travel at 4 Kroshas per hour on the Hearth, 8 Kroshas over the water and 12 Kroshas in the air.

27 Chandramukha Yantram चंद्रमुख यंत्र

 this can has its front as a lunar disk: it is dark in its central part and brightening all around. It weighs 400 Ratals. lt owns 16 doom. It brings 16,000 Ratals. It owns five floors and sixty eight cylinders. These cylinders are very useful during the saturation of the five types of smoke, of the seven powers (electric or of energy, ndr) 0f the thirty two types of powders and of the forty eight types of gas. When they are inside these cylinders, cannot cause any damages. This can can travel trough some paths dug inside the ground. It moves thanks to the help of the spirits of the wine of the 13° type. It reaches a speed of 16 Kroshas per hour. It is used in war.

28 Anthaschakra Ratham अंतचक्र रथ

 this can is modeled on the shape of a retort rod of a litter. This rod looks like the two poles angularly bent of a crusher and will be always folded on it. There are screwed 32 wheels. This cart has to be fixed on the ground. It is used to transport elephants, camels, horses, men, vehicles, etc. or to bring them from distant places. All this is done through the screws inside it. The can must be placed in the fifth circle of the war camp.

29 Panchanaala Yantram पंचनाल यंत्र   this cart is assembled by

combining five cylinders. In each of the five cylinders, there are some detailing machines. These are used to produce not only some oils, energy, etc., but also rope, powders and so on. It weighs 230 Ratals. it travels at 3 Kroshas per hour with the aids of the 9° class of the wine.

30 Tantri yantra तंत्री यंत्र

the anterior part of this cart looks like a trap of cables. Inside the cart, in its center, there is a magnetic wheel. Behind it there are some faithful representations of lions, tigers and some others wild animals, all of them made of cables. On its front there is a magnifying glass of 103 deg class Acting on the keys, these lions, tigers and so on, done on iron can be made to roar or jump up to some individuals who approach them so that no one will approach. It weights 80 Ratals. It carries a thousand of Ratals. Through the power of the energies of the 3th class of the wine it can travel at 4 Kroshas per hour. It is useful during a war.

31 Veginee Yantra वेगिनी यंत्र

this can is modeled on the shape of an umbrella. It can run very quickly turning the screws at the junction of the wheels. [t can accommodate only three people. It can travel at 8 Kroshas per hour.

32 Shaktyudgama Yantram शाक्त्यद्गुम यंत्र

it is a cart that spreads light in the sky. [t owns five floors. It contains some large glassjars (containers) in each floor. On the first floor, the glass jars are full of tar mixed with coal. On the second, the glass jars will be full of sea foam, or foam, with the extract of the pond. The jars on the third floor will be filled with the five essences of Pranaksharas's oil. Five spheres along with the mercury are mounted in those on the fifth floor. The cables of these five jars are joined according to the Shastric principles. The jars of the first floor must be filled with the electricity and through this process the containers will be loaded on the other floors. Through this trend the light can be spread in the sky. This can weights 32 Ratals. It is used to build airplanes.

33 Mandalaavarlha Yantra मंडलावर्त यंत्र

this can is modeled on the shape of a Whirligig. 1t owns six sides and sixty-four screws inside. It weights 68 Ratala. It carries 8,000 Ratals. It turns as a Whirligig around armies and crowds of people. It can spin around three times, making a distance of two crashes per hour and thanks to the help of the electrical energy and of the eleventh class of the spirit of the wine. It is useful during the war and people's mutinies.

34 Ghoshanee Yantram घोषणी यंत्र

It is modeled as a huge snake. It owns three coatings and twenty-four faces. It is filled with electricity. It owns 148 cylindrical rooms to store poisonous gasses. It can make a noise, acting through internal screws, just as thirty-two lightning. 1t releases poisonous gases while it is travelling. It is possible to hear the sound that had been made at a distance of 14 miles a quarter. The people that will be found in its vicinity will die because of the mortal effects of the deafening noise and the poisonous gases. Those that will find over eight kroshas from it, will faint. It weights 116 Rtals. It carries 6,000 Ratals. it can travel at 6 Kroshas per hour thanks to the electricity and the spirit of 13° type of the wme.

35 Ubhayamukha Yamra उभयमुख यंत्र

This machine is symmetrical on both the sides . It owns inside a fresh water stream above which rises another How of tar. In the middle there are some oils belonging to seven varieties. lt contains inside 71 keys. Working on these keys the poisonous gases, the powers or anything else that is dangerous for the life, will be wiped out in the range of twelve (about) miles around the cart and the atmosphere will be purified. lt weights 48 Ratals. It carries 108 Ratals. lt travels at 5 Kroshas per hour with the help of the electricity or the energy of the 27° class of wine. It is used to purify the atmosphere when and where you need.

36 Thridala Yantra त्रिदल यंत्र

this cart is modeled on three leaves of Bilwa patra... and has two compartments. The first is squared, the second is triangular, and the third has a hexagonal shape. Every one of these compartments own two doors. Every compartment has some Peshanee Yantras. A Peshanee Yantras is a cart that Grinds grain, such as wheat, into powder. This cart is driven by electricity.

37 Thrikuta Yantra त्रिकुट यंत्र

this cart owns two towers, as the peaks of a mountain. Each of these towers measures one hundred (bahu) yards in height. Each tower owns 32 keys inside. There are some cylinders in each key. Upon the towers there are some flags and wheels. In front there is some equipment to measure the cold. This cart reports the weather, the wind, the sun, the light, the rain, the lightning, the falling of stars and some other future phenomena.

38 Thripeetha Yantam त्रिपीठ यंत्र

this cart owns three bases. In the first there is a can with three heads as an elephant, but it owns two trunks on each head. in the second there is an apparatus with three heads; each head owns two trunks of Vyali's animal. In the third there is an apparatus with three heads, each of them has the appearance of a rhinoceros with some fangs. The three bases can be mounted together and if it is necessary can be separated. The first of these Yantras can stop the flow of the water of a river, suck its water and modify the direction of its flow. The second can break into pieces the mountains and thereby to create a passage. The third is able to dig a hole into the ground, to suck the water from below and to throw the same through the fangs posed above its head. It weighs 6,000 Ratals. It carries 80 Ratals. It travels and works thanks to the steam, for the electricity and to the energies of the 23° class. This cart is used to build bridges and streets in the water and some galleries through the mountains and rocks.

39 Vishwamukha Yantram विश्वमुख यंत्र

this cart is very spacious. There are inside twelve cylinders containing magnifying glasses. These cylinders are very big and are mounted in a way that it is possible to orient in any direction, as it may be necessary. It weighs 1.800 Ratals. It brings 40,000 Ratals. On its inside there are two floors that can be combined or separated using some buttons. It travels at 12 Yojanas thanks to the help of the spirits of the 32 typology and thanks to the steam or to the electricity. The top can be separated and can hover in the sky. By fixing to it the cylinders, it covers an area in the sky of 24 Yojana where the forests, countries, seas, towns, etc.... become clearly visible and it is possible to obtain images of the same. It is used in travels.

40 Ghaniaakaara Yantram घंटाकर्ण यंत्र

this can appears as seven Almirahs are fixed between them. In it there are several types of cables, the essence or Reavaka 0f the 16° type of magnet and also some others Dravakas are present. Inside, in each of these Almirahs, there are also two metal or white brass bells built so to produce a very disturbing sound. We can know with the waves produced all the news from the world. It is used in the collection of information and pictures.

41 Vishthrithhsya Yantram विश्वत्रिथस्य यंत्र

the cart owns a mouth completely open. It weighs 76 Ratals. It carries 120 Ratals. In front of this cart there are five buttons that look like towers. In the first tower there is a part of Chandra Kantha's stone. This stone vase, as soon as the moon rises, exudes water up to fill. The same water is used by the occupants of the cart to drink. The other towers attract the powers of the clouds, of the stars and so on. It travels at 3 Yojanas per hour thanks to the energies of the 14° class or to the electricity. It is used for travelling.

42 Kravyaada Yantram क्रव्याद यंत्र

this cart owns three sides. It weighs hundred of Ratals. It carries 10,000 Ratals. With the help of the steam can travel at 9 Yojanas per hour. It is used to travel and to transport goods. Shankhamukha Yanlram it is a cart that contains a perforated device with five faces similar to a shell called

43 Shankha Mukha Yantra षण्मुख यंत्र

There are some keys to expand or to contract the can when or where it is necessary. It weighs a thousand of Ratals. It is used to build wells, deep holes or to drill mines. [t can dig 213 bahus or yards in an hour. It is used for the purposes contained in the description.

44 Gomukka Yantra गोमुख यंत्र

this cart is modeled as the face of a cow. It weighs 80 Ratals. It carries 700 Ratals. There is a continuous flow of running water inside its mouth. It travels at 2 Yojanas per hour with the help of the energies of the 20° class. It is used for provisioning the water.

45 Ambaraasya Yantra अंबरास्य यंत्र

for the viewers this can looks like the sky. It weighs 180 Ratals. It carries 2.400 Ratals. It is used to transport elephants, camels, and so on. It travels at 3 Yojanas per hour using the electricity and the steam.

46 Sumukha Yantra सुमुख यंत्र

this cart shows the wonderful likeness of a crab It weighs 118 Ratals. It carries 1.150 Ratals. It can travel through the use of the energies of the 14° class, through the steam or the electricity. It travels at 2 Yojanas per hour on the earth, 4 Yojanas in the air and 3 Yojanas on the water. It is used to travel and to transport goods from a place to another.

47 Taramukha Yantra तारामुख यंत्र

 the spheres. which are made with the metal)mtluu [mm the falling stars. are called Taramanis. A can that contains these spheres is called 'Ihaaraamames Yantram. In its inside there are three big C) Iindrical pillars. It is also possxble to [ind inside this can a smaller (art that contains some draavakas or acid , electricity , crystals and so on. At the bottom of the three pillars above the some ehut o.ns Using the first button "1' Produce a bright light as a rainbow Instead \\ tth the second button will on a shining light as the sun when is covered by the clouds. Moving the third it mil be released a smoke similar to the dew. When this can sails on the sea it is possible to take some images of all the cans or animals that are nttmng up or under the sea stitfate. It is used to discover some obtects that are both above and below the surfaee ol the sea.

48 Manigarbha Yantra मणिगर्भ यंत्र

 this round shape. Inside this can there are some spheres called Souraka. Paavaka. and so on that attract the heat of the sun's ray 5. It "('lghb 64 Ratals. lt (antes 70.000 Ratals. It contains I2 faces to let the sun's rays to enter. It travels at 3 Kroshas per hour through the energies of the third class. It is used to travel and to capture the heat of the sun '5 rays

49 Vahinee Yantram वाहिनी यंत्र

this can owns I6 buttons and I2 metallic (\linders. It is 32 feet high [or II of cimumterence. 'lhere are under it 48 drill equipments. lhere are 96 wheels that thmw away the mud excavated. 22 lte)s that let to tltg the stones and there are 12 deviees to draw tsater. This can can be seutrely anchored to the

round. 'I he "am so sue ' [Ions like rivers. 'Ihis can can dig the ground till
32 thousand Ieet deep. It is used to dig the mound

50 Chakranga yantra चक्राङ्ग यंत्र

ths cart designed oh the shape of a trap. throughout the structure there
are some wheels with stones. Moving a wheel full of wind this one will
go out. moving another one some inner it will be released. If activated it
releases poisonous gases , colors energies etc . it travels at 2 Kroshas per
hour It is used in different wavs.

51 Chaitraka Yantra चैत्रक यंत्र

this can is modeled on the shape of a scorpion . Inside there are 34 joints.
There is a button at each joint . Each key is different numbered and
colored. Basedon the key which will be activate , will he produced music.
conversations. melodious instrumentals. images. and many other
wonders.Those who will approach will be photographed , both in their
appearance and in their mind. It is used in Bhedopaya or to conquer the
enemies with deception.

52 Chanchupata Yantam चंचुपट यंत्र

this cart has the shape of a bird with open mouth. It owns four wings.
There are five buttons in each of these wings. From its open mouth some
cables are connected to the ground. Till these cables will be connected to
the ground this will acquire a peculiar power whereby the people, if
standing in this area, will be numbed. Acting the keys that are connected to
the wings the people that will be in the infected zone they will feel weak or
the ground will crack, depending on the role that will be required for the
keys.

53 Pingaaksha Yantram पिंगाक्ष यंत्र

this cart has the appearance of a litter. This cart has all over its body many green eyes. There is a button for all these eyes. The can must be fixed firmly on the top of a mountain. It is 60 feet high for 14 for the circumference. It can also be placed in a town when this is surrounded by the enemies. The keys in this cart are made and fixed through some cables that are extending below the surface of the earth for 24 miles all around the area. inside the cart some buttons are arranged and numbered for all these extemal keys. Working on the first button we will operate in a particular key and the doors of the fort will close. Working on another the ditch will be filled with water. In this way, working on the other buttons, according to the default order it will be possible to create with each button some fantastic phenomena such as huge river of fire, flooding water, cyclones, etc. This cart is used to defend some towns or villages against stronger opponents.

54 Puruhootha Yantram पुरुहुत यंत्र

this can owns the appearance of a Mrindanga, or a musical instrument. It is 25 feet high and measures just as many feet in circumference. Inside the can there is another Shabda spota Yantra. When the button is activated a terrible noise bursts equal to the simultaneous roar of 63 lions. It used for the nature of its job.

55 Ambareesga Yantra अंबरीषग यंत्र

this can is modeled on the shape of an upside down terracotta pot. It is 46 feet in high and 23 in its circumference. It owns in each of the four sides, some keys that are looking like the feet of a turtle. It travels on the water at 6 Kroshas per hour thanks to the help of the Chakra Bhastrika.

It is used to search the things on the earth and under the surface of the sea bringing them to the light.

56 Bhadraashwa Yantram भद्राश्व यंत्र flying motorbike

its shape is modeled on that of a horse. It owns a tail about 38 feet in length. It weights 54 Ratals. It gallops as a horse with the help of the spirits of the 32° class. It has the speed of three horses. There are on the top some keys with three faces. it moves as a horse in a circular way, when it is put into operation through the keys. It covers a distance of 12 Yojanas per hour. There will be issued some bright sparks as he gallops, shining of light that will dissipate all the dew or the fog that is covering that zone and will clean all the atmosphere. It is used in the places and in the times of the dew where and when the same dew obstructs the view.

Note- Dayanand Saraswati quotes Bhoja for Motorbike with similar design

घट्यैकया क्रोशदशैकमश्वः सुकृत्रिमो गच्छति चारुगत्या ।

वायुं ददाति व्यजनं सुपुष्कलं विना मनुष्येण चलत्यजस्रम् ॥

57 Virinchi Yantram विरिंचि यंत्र

its shape is like a globe. Around it there are 32 cables of 80 feet in length and 40 of circumference, both in the front that on the back of the cart. There are three keys for these cables. Operating on the first key the cart is charged with powder and projectiles. lt shoots operating on the third key. It destroys in pieces the mountains in a measure of 24 feet for each shot. It is used to build tunnels in the mountains and among the rocks.

58 Kuladhar Yantram कुलाधार यंत्र

this cart is modeled in the figure of a crow. It owns three beaks as those of the crows. In its inside there is some electrical machinery. On its top there are some keys similar to some small boxes where there are some round buttons. When this cart is fixed on the rocks and starts to work, it extracts with the help of its beaks some slabs of stone of the desired size. It is used to cut the stones.

59 Balabhadra Yantram बलभद्र यंत्र

this cart is shaped on a metal boiler tipped. It is 64 feet long and 16 feet large. On both sides there are 16 fixed plows of 16 feet for 4 of wider Each plow owns two winds. At its beginning and its end there are some rotating screws. On its inside there is some electrical energy or a steam boiler. Upon these cart there are 24 buttons. Under these buttons there are some Wheels. On a side there are 32 screws. The can starts to plow the land when the buttons are pressed. It moves at 3 Yojanas per hour. It plows an area of 3 Yojana from 64 feet, per hour. The depth of the mud raised from the ground is about 3 feet. It is used to cultivate the land.

60 Shaalmali Yantram शाल्मली यंत्र

this can has a square shape and a white color such as that of Shireesha's acaria. On the top there are sixteen buttons, each designed for a specific purpose. Turning the first key, a pair of hands appears like the trunks of elephants that can support a weight of hundred Ratals. Acting the second button such weight will be placed wherever is needed. The other buttons are intended to lift weights from deep water, to place the pieces of stones, of wood or similar, or to arrange out of the water for bridges under construction. It can bring down weights from an altitude of 200 feet.

61 Pushpak Yantram पुष्पक यंत्र

it has the shape of the rising moon while it is forming. It is equipped with many frames that are suspended in it. There are many frames on each side and 8 of them suspended in the middle. On the right side there are some cans similar to pigs, while on the left side there are some carts to see. In the middle there are some wheels screwed suspended by the chains. There are also two wheels. This cart can be placed in a place where the wood will be cut and sewn. If the first buttons of the top wheel are activated, the above mentioned cart similar to pigs will descend one by one. Activating the second one, the cart will fall on the trunk of the tree and will shake it, and will cut it with a tremendous noise producing a great amount of smoke and lire that spreading around for 16 miles, burns all the waste on the soil and cleans the air. By the action of the fire in the trees, the oils etc. will be extracted and stoned in containers at the base of the same trees. The heat of the fumes produced by the fire makes all the trees in the area soft as plane trees. The leaves fall. Operating on the third key some cans descend and wonder in the place exhaling tremendous breaths. Thanks to these strong winds the dust of such area will be swept away and the land will be deforested. In the same way, if we will operate the button on the left side, from their cribs the shows will drop down one by one. Tuming the first screw of this wheel the saws can arrive ready on the side where the trees will be cut. Operating on the third screw, the saws will come back to their place and a pair of hands, as trunks of elephants, will descend. The pairs of hands will reap the pieces of wood that have been seen. This can weights 180 Ratals. It can move into the forest thanks to the power of the steam.

This cart had to be fixed on the ground. It can saws 3,200 Ratals of wood per hour. It used to saw and to cut large quantities of wood.

62 Ashtadla Yantram अष्टदल यंत्र

it is a machine modeled on the shape of a lotus containing 8 petals. Under each of these petals you will find some containers. In each of these containers you will find eight things that are smoke, light, water, steam, air, Rushakam, Vishasaram, Manjusham Katusaram that are described in the Meghotpati Prakaranam. In the center of the lotus there is a key, in which there are eight screws for the eight petals. By actuating each screw what that is connected to the petal will be brought to the top and will form a cloud. Pressing the central button of the steam it will be irradiated as the sunlight. As soon as the heat of these fumes will act on the clouds previously formed, it will stan to rain. This machine is specialized in producing the rain.

63 Souryayana Yanlram सौरयान यंत्र

 this machine looks like a column of 116 feet in height and 58 in circumference. At the top there is a sieve containing some holes and it is made of glass of the 96th class called Somapa. From this sieve, contained in the pillar, there are twelve machines in order and above the sieve there is a glass cover of the 97th class called Somasya Darpana. Above this there is a wheel with a glass cover called Kumudinee containing some rays made of glass class 98th and called Chandrika Darpana. In the twelve points of the machine, there are twelve screws above and twelve lower screws. Turning the screw the contents of the machine will rise in the proportions as for electricity, cold fluid, Shaitya Drava, Sudha Mushee, Soonruta, Pushkalee, Pranada, Dravinaamrutha, Sooraneee, Jambaalee, lulita, Vaachaklavee, Gacyoosha. Through some cylindrical tubes that are attached to the wheels of the sieve these energies will flow touching the glass on the top cover. Turning the screw for electricity the wheel starts to urn 1192 RPM. 50 the energy of the rays of the moon called Someeya will be attracted to this wheel and fall through the sieve.

Thus, this energy will fill the container below in the form of gas that must be kept tight. Its use is that, when the limbs such as the head, hands or feet, of a person are cut will be resettled in the right place and the body will be kept in a special container. The body should be wrapped in a blanket of bark of a plant called Vaarshneeka Valkala.

When to that body will be injected with the aforementioned gas Somadrava for 5 Rajanikas, the body will be resurrected. This must be done within five minutes after the wound was caused. It is used to put in place the limbs immediately, or resurrect those killed in the manner described above.

Chapter II

Aerial Vechicles of Aryans & their mysterious transfer to NAZIs

At present age we have very few works on ancient technology . Out of them most were dictated by Pandit Subbaraya Shastry in early 20th century . These works get quoted in works of Sri Talapade & Sri K V Vaze , former who was doing research in aeroplanes & later was overall researcher of ancient technology. We list below types of these aerial vehicles from above works . We will trace them to their modern counterparts & the routes via they went to Germany & then to US from India in early 20th century.

From Yantra Sarvasva , most ancient sanskrit book on machines :

There are three works which bear the author as sage Bharadwaja viz Akasha Tantra , Amshu Bodhini & Yantra Sarvasva .
Yantra Sarvasva is divided into 8 major chapters or adhyayas which describe following machines or vehicles related with transportation

340 types of Land mobiles
783 types of Sea vehicles
101 types of Aeroplanes

From notes of Sri K V Vaze we get an index of subchapters of Yantra Sarvasva called adhikaranas .
From 66th to 73th adhikarana in chapter 6 ,

In Sri Talapade's work we get very little information about above vehicles of eight types . In published versions of Vaimanika Shastra (which seems to be compilation of different adhikaranas(chapters) of Yantra Sarvasva & other works) we see description of Shakuna, Sundara , Rukma & Tripura vimana . These Vimanas except Tripura do not have separate adhikarana . They come under जात्याधिकरण of Yantra Sarvasva . Only Tripura Vimana has its own separate adhikaran 77th one in chapter 7th . Probably as Tripura vimana is multi floor aeroplane , it is put under separate chapter of similar vehicles . Other vehicles in 7th chapter are related to interplanetary travels as some adhikarana like 76th & 81th hint . 81Th adhikarana is named मांडलिकाधिकरण. Here we should remember Mandalika is name of planet or graha mentioned in Vaimanika Shastra . probably is the name of earth's moon .

Shakuna, Sundara & Rukma vimanas work majorly on air & oil as fuel . They can be grouped under Bhutavaha Vimanas of 2nd type .

Classification

	Type	Fuel	Names given by Sage Bharadwaj	Present similar vehicles
1	शक्त्युद्गम Shaktyudgama	Electricity	Shakuna, Sundara	Airships like Zeppline, Solar Impulse 1,2
2	सूतवाह ** Bhutavaaha	Mercury	Rukma, Tripura	TR 3B astra , Die Glocke or The Bell
3	धूमयान Dhumayaana	Exhausting gases formed by steam		Not existent
4	शिखोद्गम Shikhogama	Oil		Most of the all current aeroplanes
5	अंशुवाह Amshuvaha	Solar rays		None This power is different than just photoelectric

6	तारामुख Taramukha	From magnetic properties of meteoroids		current UFOs research in Germany & Australia
7	मणिवाह Manivaha	Electricity from Air in higher atmosphere		Not existent
8	मरुत्सखा Marutsakha	Vortex flow	Rukma ***	Haunebu , Vril series of Germany

** In some published versions it is भूतवाह meaning five elements as fuel

*** Rukma aircraft can be worked on air , water or mercury as fuel depending upon required speeds & distance or dimension travel as we will see below . Shri Talapade , aeronautical enthusiastic & researcher from India was working on aeroplane on Marutsakha type which was based on mercury as fuel . He was developing such aircraft under guidance from Shri Subbaraya Shastri . Success or failure of his experiments is not known .

Construction & Working similarities between Rukma Vimana & Haunebu

As stated in Vaimanika Shastra , outer body of Rukma Vimana should be in shape of Kurma ie tortoise shell . We are not going under the matter that how Vaimanika Shastra technology went into Vril Society's hands or SS hands actually . But we can observe similarity .

Working principle Similarities

On working principle of Rukma Vimana, its given in Vaimanika Shastra following description .
Lalla gives the form of ayaschakra-pinda:
12 feet long and wide, and 8 kankushtas in weight, they should be made round like a
grind-stone. They should be inserted in the beaks at the 8 centres. From each chakrapinda
up to the electrical generator chain wires should be connected with switches.
Batinikaa-Stambha

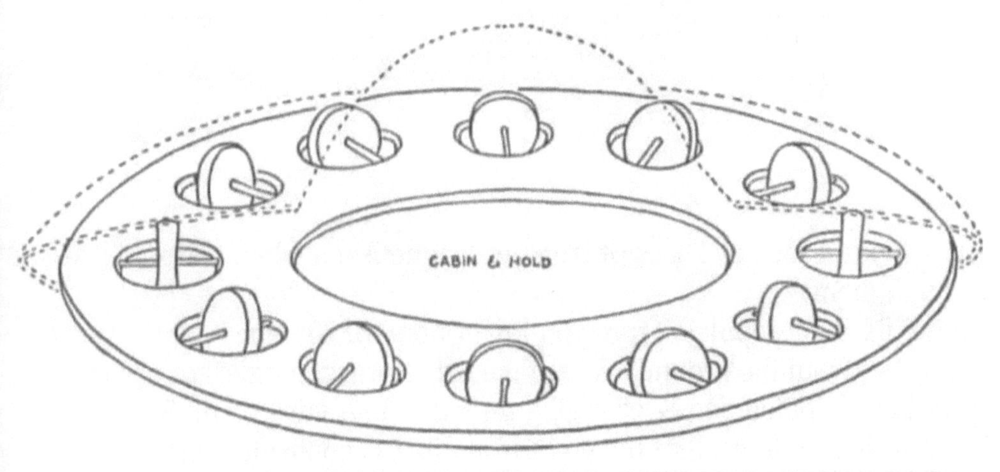

Now see design construction of Haunebu from SS vaults also called Mark 2 in design documents

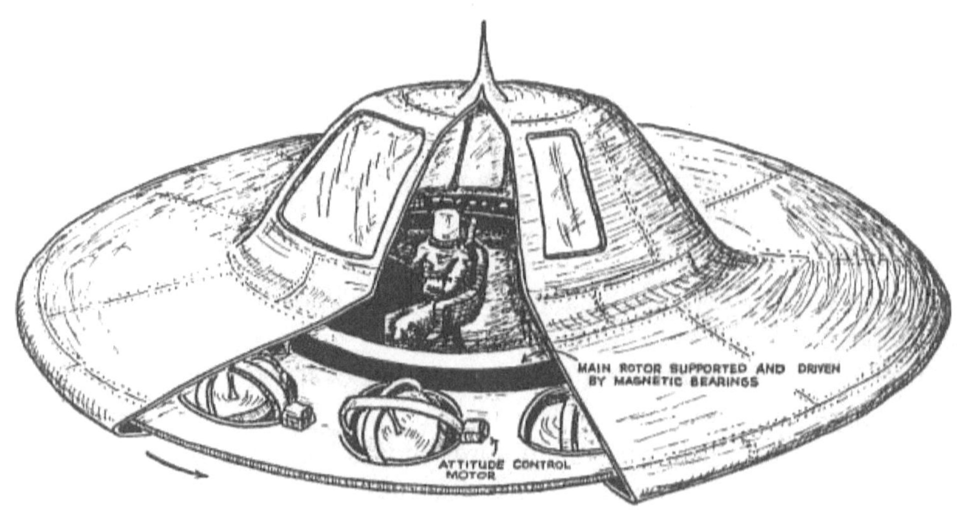

MARK II FLYING SAUCER

ELECTRONIC CENTRIFUGES BASED ON THE VORTEX DRIVE ARE MOUNTED IN GIMBALS TO TURN IN SYNCH WITH THE REVOLUTIONS OF THE MAIN ROTOR DISC.
THE TUNED ELECTROMAGNETIC FIELD GENERATED BY THE VORTEX DRIVE CAUSES THE VEHICLE TO BE CARRIED BY THE EARTH'S ELECTROMAGNETIC FIELD LIKE A DIRIGIBLE ELECTRON.
CONTROLLED GEOMAGNETIC PROPULSION IMPROVES THE DESIGN EFFICIENCY TO THE MARK III STAGE.

Flying principle

Ayaschakra Pinda or Discs of Iron get magnetiszed when current is passed through them .
These discs put repulsive force on lower body of Vimana resulting it taking off from the ground . After take off , gyroscopic design of these discs help in balancing & changing direction in 45 deg on all 3 axis .
Electric supply to each of the eight iron discs is controlled through different switches & helical gear mechanism is used for various switching operations .

One such gyroscopic drive was named as Thule Tachometer engine in German secret weapon construction .

Speed of the Rukma Vimana

Generating 25000 linkas speed, which will give the vimaana 105 krosa or nearly 250 miles speed per ghatika, or 24 minutes.

This is around 700 miles per hour . Note RPM is 25000 , depending upon gear mechanics & linear or rotary propulsion type , speed can be increased much higher .

Lets compare it with Haunebu speeds . Remember this Rukma Vimana is possibly the first version of Haunebu Aircraft . Haunebu III aircrafts was having speed of 6000km per hr & in later versions or models speeds of 21000km per hr were possible .

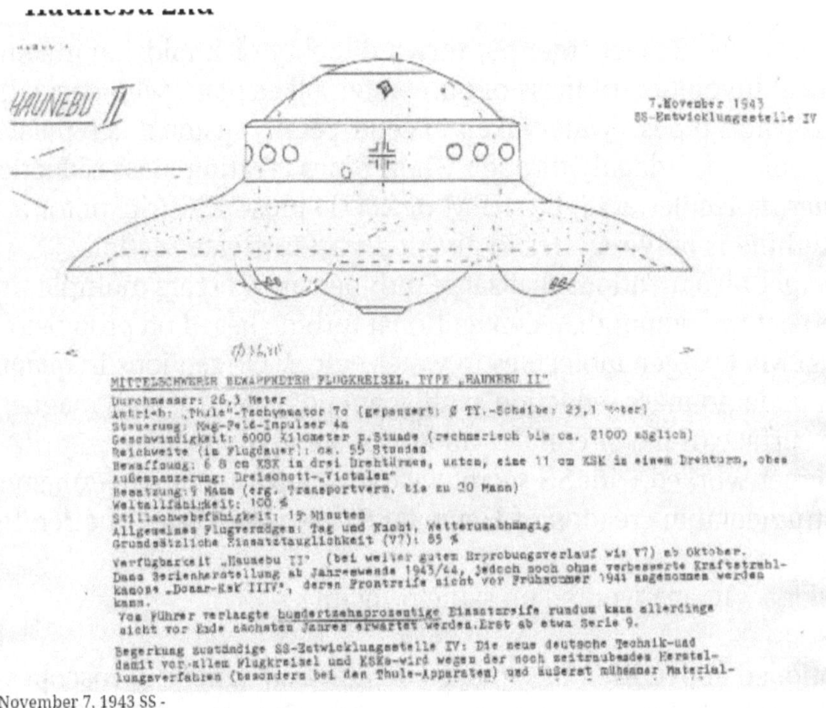

November 7, 1943 SS -

Cavity Propulsion & Mass less travel

As per US Patent 10144532B2 named Craft Using an Inertial Mass Reducing Device, patent assigned to US navy on Dec 4 , 2018 , Space or Water propulsion through quantum vacuum created cavity without any resistance of air,water,ions or gravity .

This technology creates quantum vacuum outside the aircraft by electromagnetically induced mass reduction & polarization of outside vacuum by microwaves resulting in negative pressure or repulsive gravity In such system aircraft is moved without any resistance & is capable of achieving speeds higher than speed of light depending upon mass reduction percentage .

Such aircraft should travel through water , air or space without resistance , but in maximized work condition of quantum vacuum surrounding the aircraft, it should travel through solid matter also without any resistance . This though looks notorious but possible on technology mentioned in said patent .

Viktor Schauberger's IMPLOSION machine

German scientist Schauberger put forward theory of Implosion in which mechanical involution of mass of air / water takes place with specially designed whorl pipes . Water taken in centripetally inside this repulsive turbine cools off suddenly through whorl pipes creating mass reduction in air / water molecules at higher RPM of 20000 plus . Electric motor at axis of this turbine is only requires to give it a start to create suction . Schauberger also mentions that same turbine can generate multiple times (9 times) energy output than conventional turbine based on propulsion . At higher RPMs Oxygen molecules in water / air or Origen ions in quantum vacuum undergo mass reduction as they are polarised by high magnetic field on surface of the aircraft by microwave emissions . Schauberger worked with SS secret weapons department in Germany & was instrumental in creation of Haunebu & later Vril series of aircrafts .

Was Rukma Vimana worked on same principle ?

As mentioned above Rukma Vimana had electromagnetic gyroscopes , 8 nos on 8 cardinal directions under the vimana disc . Vimana also had multiple pipes having spiral designs going from centre . Central axis of Vimana had suction pipe on which electric (hydraulic) motor was mounted . This is strikingly similar to design of Haunebu aircrafts designed by Schauberger .

Here is how his designed aircraft looked . See its resemblance to tortoise shaped outer structure given in Vimana Shastra for Rukma aircraft .

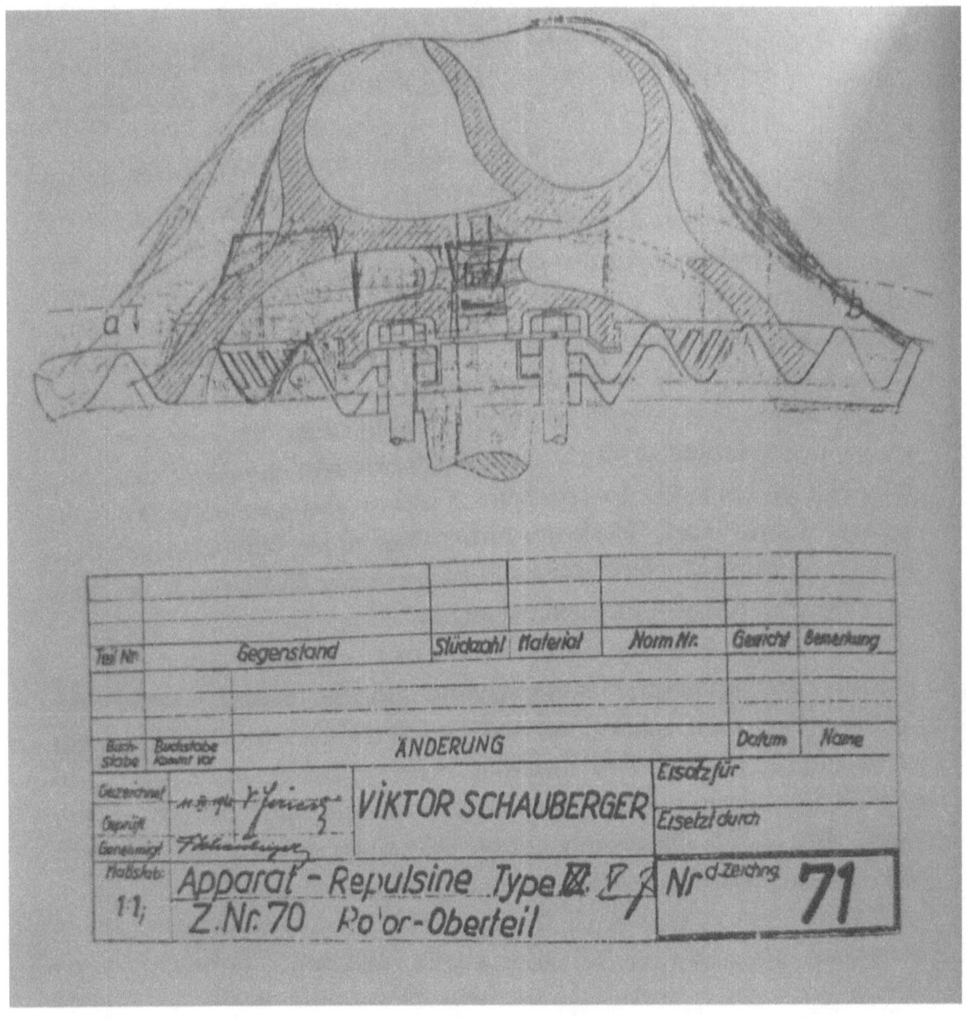

We mentioned above the US Patent granted to US navy in Dec 2018 which is on same technology of quantum cavity propulsion . Designs suggested in that patent are flat triangle or rectangle shaped . In 70 years change from disc to triangle we can call it great development right ? Interestingly same triangular design is connected with secret US air/water spacecraft TR3B astra seen & recorded at many war zones in Afganistan & Iraq. Also many times misunderstood as triangular UFO by many UFO witnesses . There are also instances of commercial & defence pilots witnessing this triangular shaped aircraft taking sudden 90 deg turns . We also should note similar aircraft Aurora in 1980s developed in US on similar technology .

Aurora aircraft Model in below picture

Curious eye can notice that in last 90 years history of this anti gravity or quantum vacuum
propulsion drive , same machine is mentioned under different names & said to be working under different technologies . From Pulsating device or Implosion machine from 1930, to magnetic field disrupter in 1980s and now to Internal mass reduction device . Purposeful mischief has been played by certain groups to mislead general masses from such technology & hide the real mechanism & source of such technology .

From India to US via Germany

NASA's Dragon spacecraft designed exactly like Rukma Vimana

Here is design of Rukma Vimana prepared by Subbaraya Shastri

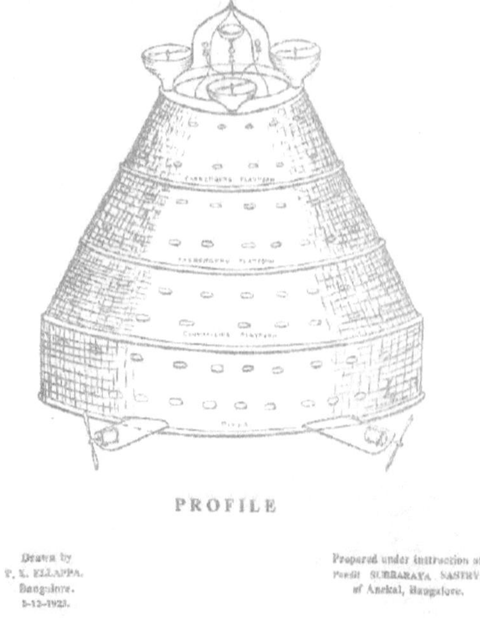

Bharadwaj Vimanashastra & Akasha Tantra were known to Subbaraya Shastri from 1885 who published articles on ancient vedic sciences in monthly magazine Bhautika Kala Nidhi from 1915 . Magazine was published by B Suryanarayan Rao in Chennai (Madras) in India .
Various articles of Subbaraya Shastri, his published works Vaimanika Shastra , Prasthana Trayee mention many ancient works on aeronautics & aviation , naval technology .

Chapter III

Scientific Works of Sage Shaunaka

In ancient sanskrit literature , we find multiple personalities having name Shaunaka .
One who was preceptor of one branch of Atharvaveda & one who wrote index work on Rigveda with its grammer are prominanent among them . Shaunaka sage earlier belonged to Bharadwaja Kula (Bharadwaja are Angirasas). Later he became of Bhrigu clan . Thus he inherited best of knowledge of both streams of advanced technology . Bharadwajas were expert in Metallurgy & Machines while Bhrigus were experts in Construction engineering & Weapons . As Rigveda is Dhatuveda ie Mechanical engineering & Atharvaveda is Civil enginnering or Shilpaveda . As expert in both fields he wrote on Aircraft manufacturing , Artifical gems making , Astrophysics & Astrochemistry (see below) . He also created book on Shilpa Shastra as per Matsya Purana , this book is not available now .
Shaunaka is also called as Kulapati in Mahabharata . Head of University who has 12000 students under him is called Kulapati (Aranyaka 1.1) .
He did 12 yrs long sacrifice (Yagya Satra)in Naimisharanya (present north eastern UP state in India)after 1000 years from start of Kaliyuga . During his sacrifice he dictated currunt available versions of Vedas & Mahabharata to sage present there . Later he withdrew to Kalapa Khsetra with 12000 sages with all higher scientific works on Veda .

अदिनिक्ष्मावाय्वर्केन्द्वमृताम्बरशक्तयस्सप्त वैमानिका इति ता नामान्यनुक्रमिष्याम। उद्गमा पञ्जरा नूर्यशक्त्यपकर्षिणी विद्युद्द्वादशका परशक्त्यपकर्षिणी कुण्टिणी मूलशक्तिश्चेति॥
अथ विमानेषु त्रेताया पञ्चविंशतिस्ते मान्त्रिकास्तेषां नामान्यनुक्रमिष्याम।
ुष्पकाजमुखभ्राजस्वज्योतिर्मुखकौशिकभीष्मशेषवज्राङ्गदैवत-
ज्वलकोलाहलार्चिषभूष्णुसोमांकपञ्चवर्णषण्मुखपञ्चबाणमयूरशङ्करत्रिपुरवसुहार-
ञ्चाननाम्बरीषत्रिणेत्रभैरुण्डा इति॥
द्वापरे तान्त्रिकाष्षट्पञ्चशत्तेषा नामान्यनुक्रमिष्याम। भैरव-नन्दन-वटुक-विरिञ्चि-तुम्बर-वैनतेय-भेरुण्ड-
नकरध्वज-शृङ्गाटकाम्बरीषषास्यसैहिक-मातृक-भ्राज-पैङ्गल-टिट्टिभ-प्रमथ-भूर्ष्णि-चम्पक-द्रौणिक-

रुक्म-पुङ्खभ्रामणि-ककुभ-कालभैरव-जम्बुक-गिरीश-गरुडास्य-गजास्य-वसुदेव-शूरसेन-वीरबाहु बृसुण्ड-
गण्डक-शुकतुण्ड-कुमुद-क्रौञ्चिकाजगर-पञ्चदल-चुम्बुक-न्दुभिरम्बरास्य-मायूरक-भीरु-नलिककाम-
पालगण्डर्क्ष-पारियात्र-शकुन्त-विमण्डन-व्याघ्रमुखिविष्णुरथ-सौवर्णिकमृड-दम्भोलि-बृहत्कुञ्जमहानट इति
अथ तिष्ये कृतकभेदा पञ्चविंशतिस्तेषा नामान्यनुक्रमिष्याम॥
शकुना-सुन्दर-रुक्म-मण्डल-वक्रतुण्ड-भद्रक-रुचक-वैराज-भास्कर-गजावर्त-पौष्कल-विरञ्चि-नन्दक-
कुमुद-मन्द-रहस-शुकास्य-सौमक-क्रौञ्चक-पद्मक-सैहिक-पञ्चबाण और्यायण-पुष्कर-कोदण्डा इति॥
अथाकाशमार्गाण्यनुक्रमिष्यामो। रेखामण्डलकक्ष्यशक्तिकेन्द्रभेदाद्
भूतशक्तिप्रवाहमार्गाण्याकूर्मादावारुणान्तं
वाणमवष्टभ्यैकचत्वारिंशत्कोदयैकपञ्चाशल्लक्षनवसहस्राष्टशतसंख्याकानि भवन्ति तेषु भूरादि
सप्तलोकविमानास्सञ्चरन्ति इति॥

तदुक्तं शौनकीये-

प्रवाहद्वयसंसर्गादावर्तनमिति तान्यनुक्रमिष्यामः। रेखापथे शक्त्यावर्तनं मण्डले वातावर्तनं कक्ष्ये
किरणावर्तनं शक्तिपथे शैत्यावर्तनं केन्द्रे घर्षणावर्तनमित्यावर्ताः पञ्चधा भवन्ति इति। <u>आवर्ताः पञ्चसु
पञ्चेति हि ब्राह्मणम्</u>॥

२. तदुक्तं शौनकीये –

अथ तृतीयवर्गस्थाभ्रकनामान्यनुक्रमिष्यामो शारदङिकलसोममार्जालिकरक्त-मुखविनाशका इति।
सोमेनैवैतदिति केचित्॥

तदुक्तं शौनकीये-
िूत्र- अथ िज्ञान् प्रिक्ष्यामः। खवनजाः कुर्जाः वशर्जाः कृतकाश्चेवत। तेषां ित्िरो िणाल
ब्रह्मक्षवत्रयियैश्यशूरभेदात् पुनः प्रवतिणालत्पञ्चसिशवतः पञ्चसिशवतष्षि् सिशच्छूरतयाहृत्य
पञ्चोरशतभेदा िज्ञाणाववमवत॥

Sage Shaunaka wrote extensive works on Metallurgy & classfication of
elemetns , Gemology . He gave methods of creating artificial glass,
gemstones , diamonds for industrial & space travel use . His vast book
Shaunaka Sutra is example of his wide scope into things . Quotes from his
books frequent the pages of Vimanashastra & Amshubodhini , top most
technolgical works of Aryans

Chapter IV

Shilpashastra – Aryan Engineering

Engineering is a material branch of human activities and in its study one get an eyesight into metaphysical and spiritual conceptions of ancient Indians. Many incongruities –seemingly absurd manners and customs of Indian communities are rationally explained by a study of physical sciences and their worldly developments by ancient Indians. By the kind permission of the editors of this volume , I shall in these new pages endeavor to show how valuable and useful the study of these sciences is. I shall   feel myself very fortunate , if I am able to draw the attention of enlightened public to these stories of knowledge. I have been, for last fifteen years, trying to get these works introduced into the Indian colleges and universities, but so strong is the perversity of sight created by Western studies, that I have   not even being kindly heard, much less duly appreciated. The Sanskrit name for engineering is 'Shilpa' The very word 'Engineering ' in English means ' a science dealing with engines', but the word 'Shilpa' has much wider significance, it is derived from root 'Sheel' to satisfy and means a science dealing with the production of things required for satisfaction of human wants', in short , Shilpa deals with the production of all amenities. All things from cooking of food to the construction of forts and towns, from the manufacture of a pin to the building of a large mill, from a field terrace or the building of a hut to the construction of harbor or of magnificent palaces are included in this term. Shilpasamhita -Engineering encyclopedia is divided into three   parts according to the labor required.

> धातुनां साधनांच वास्तुनां शिल्पसंज्ञीतं ॥
> भृगुसंहिता

- Dhatu – the part dealing with extraction of materials.
- Saadhan- the part dealing with the conveyance of materials and
- Vastu –the part dealing with the construction of structures.

Extraction is easy, conveyance is more difficult and construction or manufacture is the most difficult of all. These parts are again divided into three sciences such as under:-

कृषीर्जल खनिश्चेति धातुखंडं त्रिधाभिदं॥
नौकारथाग्नियानानां कृति साधनमुच्चते ॥
वेश्मप्राकारनगररचना वास्तुसंज्ञितं ॥
भृगुसंहिता अ.१

Dhatukhanda is sub-divided into Krushi –Agriculture, Jala-Hydraulics and Khani –Mining. Saadhan Khand is subdivided into Nauka- naval architecture Water transport, Ratha- roads and surface transport and Agiyan-Aeronautics and balloons or Air transport/ Each of these sciences is also defined.

वृक्षादिप्रसवारोपपालनादिक्रिया कृषिः ॥
भृगुसंहिता अ.१

Krushi- agriculture is the science of production, regarding and uses of trees , animals and man. All things in the universe are divided by Indian botanists into two parts viz. Inorganic(Anshan) and Organic(Sashan). Organic things are again subdivided into three classes according to the flow by which they are fed; viz.

- Urdhwa strot- Things having upward flow,
- Tiryak Strot- Things having an oblique flow,
- Arwak Strot- Things having a downward flow,

The Urdhwa strot things are divided into six types viz. Vanaspati, Aushadhi, lataa, twakasaar, Veerudh and Drum. This is Botany of modern times. All animal life takes food by mouth and throws the residue at the tail. This flow is thus in a slanting direction. All animals are included in Tiryak strot. This is biology of modern times. All human life is called Arwak Strot as they take their food through mouth situated high in the head and throws the residue through apertures vertically down the trunk. The flow is thus downward.

Krushishastra or Agriculture deals with production, bringing up and use of all the organic beings. Thus the science includes modern Botany: Biology and Sociology. An engineer in his professional work has to deal with both organic and inorganic substances and amongst organic substances not only with the vegetable and animal kingdoms but also with human beings in all their grades. Engineers have to use all laborers, artisans and even specialists in various sciences in construction of their works and therefore must know the properties, qualities and defects of these human beings as well as the inorganic substances or of vegetable and animal substances. This branch is the study is totally neglected in Western engineering and engineers and managers of large mills and factories are left to do the best they can in the matter without a proper scientific study of the materials (human nature) they have to use. An engineer must know the qualities of a philosopher, of an electrician, or of a carpenter as well as the properties of wood and other things he uses for his work.

> संसेचन संहरणं जलानां स्तम्भनं जलम ।

Jalashastra or Hydraulics deal with three subjects viz.
(1)Water supply or Irrigation (Sanchetan)
(2) Drainage of water or reclamation (SanharaN)
(3) Storage of water or Bunding (Stambhan)
Indian engineers are known to be born irrigation engineers as is manifest from the various small and large canals and other irrigation works built by them in Sind, Deccan and other parts of India from times immemorial. Indian engineers say that the flow of water in a river is Matsyamukh and Kurmaprushtha, i.e. the head of water is like the head of a fish , the portion in the centre having greatest velocity flows front most and the surface of flowing water is like the back of a tortoise ,the portion in the centre with the greater velocity being highest with sides falling towards both the banks:

> पाषाणधात्वादिद्दतिस्तद् भस्मीकरणं तथा ।
> धातुसांकर्यपार्थक्य करणादिक्रिया खनि: ॥

Khani Shastra or Mining science deals with;
(1)Druti –Quarrying of stones. and metal ores.
(2) BhasmikaraN- Burning of stones/ bricks and extraction of metals.
(3) Sankar-Preparation or mixtures and alloys.

(4) Parthakya – Analysis or separation of the component parts of a mixture or alloy. The graduation of these sciences is according to the labor involved in these operations; the production and rearing of animate beings are easy and deal with the surface of the earth only; storing and removal of water are more difficult as they have to go deep into the crust of the earth in digging wells and obtaining impermeable strata for foundations; and mining and metallurgy are still more difficult as they have to use not only deep shafting and vast calcinations, but chemical reagents and forces.

नौकारथाग्निनियानानां
कृति:साधनमुच्चते ।

Transportation by running water is the easiest and so Naukashastra (Water transport) is divided into three parts viz. (1) Rafts (Tatee), (2) Boating (Nau) and (3) Shipping (Nauka). Rafts are floated on water and go as it carries them. Boat (Nau) is that helped by wind in a sail and Ship (Nauka) is rowed, sailed and floated.

पवमान सोमाभ्या नीता नौ:

The modern steamers are ships rowed by machines worked by steam or such other power.

Ratha shastra deals with construction of paths (Path), cleared roads (Adhv) and Paved or metalled roads (Rathya). In constructing roads there are three classes viz.

(1) Ghantapath- Hill roads on which the use of bells (Ghanta) was compulsory as a warning to other carts ; these had their surfaces all sloping towards the hill.

(2) Veethi- Town roads are whose surface was low in the centre and carried rain water away from the plinth of houses.

(3) Rathya –Plain roads whose surface was high in the centre.

Tunnels (Vivar) were not quite unknown though ordinary roads did not require them very frequently.

Ship (Nauka is called Jalayan or a conveyance passing over water , Rath is called Bhumeeyan or a conveyance passing over land and similarly Viman is a balloon is called Agniyan or Vyomayan or a conveyance passing over fire or through air. The word Agniyan when occurring in ancient texts is frequently misinterpreted as meaning a railway; but a text of Agastya Samhita gives the detailed construction of Agniyan is much

imilar to a Zeppelin ,the motive power instead of a motor engine , being
 flock of birds

> गरुद् मद्हंसै: कंकालैरन्यै: पक्षिगणैरपि ।
> आकाशे बाह्येद्यान विमानमिति संज्ञितम् ॥
> अगस्त्यसंहिता

age Agastya in his text gives methods for training birds and using them
or carrying balloons through the air. The want of light engines prevented
he general use of these balloons. The names of Garudavahan for Vishnu,
Iansavahan for Bramha and Grudhravahan for Agni indicates the power
hey used for propulsion of their balloons. The havoc playd by Ravana and
haalba with their balloons made the Indians detest these conveyances and
heir use was probay prohibited by common consent. In the Ramayana we
ead of Rama resigning the Pushpak Viman to Himalayas which is a store
ouse of all the best things and beings of Indian culture.

> वेश्म प्राकारनगररचना
> वास्तुसंज्ञितम्।

astukhand comprises of three science viz. Building construction
Veshma), Defense works or construction of forts (Praakaar) and Town
lanning (Nagar Rachana). The term Vaastu means something belonging to
 receptacle (Vaastu). A thing in which gods (Parmatma) resides is a
aastu and a thing that is made of vastu is Vaastu as explained in the
ollowing verse:-

> प्रासादादीनि वास्तूनि वस्तुत्वाद्
> वस्तुसंश्रयाद्।
> वस्तुभिर्निमितत्वाच्च प्रोक्तान्येवं
> पुरातनै: ॥

The great peculiarity of Indian building is that they consist of open
courtyards. The town excavated near Monenjo-Daro in Larkana district in
Sindh and great cities of Vijayanagar and those founded by other Indian
kings consist of houses having one or more open court-yards. The Raste

wada in Poona is the best example of a palace on a small scale as described in the Ramayana and Indian engineering books. The front gate of Shanwarwada is exactly described in Arthashastra of Kautilya.The temples in Nasik are exactly as required by the three Samhitas of Engineering.

- Naroshankar temple according to Kashyapa Samhita.
- Sunder Narayan temple according to Bhrugu Samhita.
- Kala Ram temple according to Maya Samhita.

This shows that up to the advent of the British rule the ancient engineering texts were consulted and adhered to. During the Maratha period the Mohemddian style was abandoned and attempts were made to revive the ancient texts and to follow them in detail as it is seen in the houses, temples, towns, drainage works at Wai, Satara, Poona and Nasik. The palaces particularly the Darbar hall at Satara is just as they should be according to ancient engineering texts.

Praakaar Shastra deals with fortifications and various kinds of defensive works. In the part of Engineering texts five Dhanurvedas are mentioned which are authored by Vashistha, Vishwamitra, Ushanaus , Jamadagni and Bharadwaja. Out of these the first four are at present available and a short description of these texts is given below:

(1) Vashistha Dhanurved – This consists chiefly the gymnastics only and this was followed by Sugreev and his army as narrated in the Ramayana. Weapons were unknown at that time and the name Dhanurved comes from breaking of enemy's body by bending and not from bow. After quarrel between Vashistha and Vishwamitra the latter invented weapons and bows.

 (2) Vishwamitra dhanurved gives the details of weapons invented by Vishwamitra who afterwards imparted his knowledge to Rama.

 (3) Ushanus dhanurved describes the formation and properties of Vyuha or the arrangement of the army. A particular formation (Vyuha) is to be used under particular circumstances and with a particular object. The art lies in the formation and development of it by which arrangements of fighters are least exposed to an attack and this is called Sanjeevan vidya, the act of counter acting the losses by deaths in the battle.

 (4) Jamadagni dhanurved describes the explosives discovered by Jamadagni and known after him as Jamadagni -living fire. Jamadagni describes some missiles(Astras) which are some sort of bombs to be

attached to arrow and exploded on striking against enemy's body. These were used by Parshuram, Bhishma and others.

(5) Bharadwaj Dhanurveda was the text taught by Dronacharya and used in the great battle of Kurukshetra between the Kauravas and Pandawas. No text of thid Dhanurved is yet found.

It was Kashyapa who invented the means of protecting human bodies against the weapons of the enemy such as shields, armors and this idea was developed into ramparts, trenches and other defensive works.

यत्ते शिल्पं कश्यपरोचनावद् ।
इंद्रियावत्पुष्कलं चित्रभानुम् ॥
यस्मिन सूर्या अर्पिताः सप्तसाकं ।
तस्मिन्ना जानमधिविश्रयेयम् ॥

This verse from Taittariya Bramhana describes the protective works used by Kashya and this in course of time has developed into Praakaar shastra (science of Forts and Castles)

In Nagar Rachana shastra the planning of towns of various shapes and sizes under different conditions is described in detail, selection of sites, supply of amenities, construction of temples, courts, palaces , recreation grounds, educational and charitable institutions , disposal of waste products and allotments of land for bazaar and trade centers are the chief sections of this science.

All town management is carried out by representatives of the various communities living in their own wards and the Mayor is called Adhipati. Kautilya in his book Arthashastra describes the formation of a town, the settlement of a country and management of a community. Poona, Nagpur, Gwalior, Indore are examples of towns planned and settled during the Maratha period. According to the texts water supply to a town was not to be through pipe-connections but through cisterns for each ward or locality. The quantity of water required for each locality is calculated at nine kumhas for Brahmin, seven for Khyatriya, five for Vaishya and three for a Shudra and one for low caste Criminal out laws (Chandalas).

In ancient literature there is no separate Yantrashastra but the various implements and machines are described with the operations where they are used. In later periods however Yantrashastra is separately treated. Indian authors divide the whole world into five elements recognized by five senses viz.(1) Aakash is recognized by ear through its property of sound. (2)Vayu is recognized by the sense of touch through its property of touch, (3) Tej is recognized by the eye through its property of color and form, (4) Aap is recognized by taste through its property of dissolving and (5) Pruthwi is recognized by the sense of smell through its property of odour. The five senses of the human body are the only means by which the existence of any thing is perceived and hence the primary elements cannot be more than five.

In Mechanics (Yantrashastra) there are five mechanical appliances suited to the operation s to be carried out as under:-

Principal elements					
Name of element		Mechanical appliance		Operation to be done	
पृथ्वी	Solids	दण्ड	Lever	उच्चाटन	Breaking
आप	Liquids	चक्र	Wheel	वशीकरण	Leading
वायु	Gases	दंत	Gear	स्तंभन	stopping
तेजस	Light	सरणि	Inclined plane	जारण	Concentration
आकाश	Ether	भ्रम	Screw	मारण	Annihilation

दण्डैश्चक्रैश्च दन्तैश्च सरणि
भ्रमकादिभि:।
शक्ते: संवर्धनं किवा चालनं यंत्रमुच्चते॥
यंत्रार्णव

Yantra is a contrivance in which one or more of mechanical appliances noted above are used for transmitting and changing direction and place. Power used is called Beej, outturn is called Shakti and fulcrum is called Keelak. From the above details it will be seen that ancient Indians were not ignorant of anything except the steam or motor engine. Coal and petrol was unknown in India and engines propelled by these agencies were consequently unknown. Indian engineers however used vegetable and animal products very profusely and this is not a small legacy. Mineral products, however extensive are sure to be exhausted one day or other, whenever the day might come and Indian engineers rely more upon the everlasting sources of vegetable and animal products.

Having so far described in detail the extent of the knowledge of engineering in India, we shall now turn to some of the peculiarities of engineering practices in India under three heads viz. (1) Theory (2)Design and (3) Execution.

(1)Theory

In Indian Engineering, there are certain first principles which are to be followed in all works.

(1) The most important is the axiom that the effect produced is proportional to the power applied/ Money, time and energy are the principal powers used in engineering works and nothing can be done without utilizing some of these powers.

(2) Things are to be used in their natural state, for example, in case of a post. Its bottom must be the bottom of the tree and its top, the top of tree of which it is made. All wood is to be used with the bottom down.

(3) Time is changing and engineering works must provide for this change; for example a house should be warm in cold weather and cool in hot weather or a ship should float well not only when it is loaded but also when it is empty.

(4) Previous experience should always be duly considered and in making alterations, the changes in disposition of forces under the altered conditions should be taken into consideration

> वर्णगंधरसाकार दिक्शब्दस्पर्शनैस्तथा ।
> परिक्ष्यैव यथायोग्यं संग्राह्यं वस्तुमात्रकम् ॥

An engineer must know the color, smell, taste, shape, crystallization, sound and touch of all things and he should select the best material available. In order to import this knowledge, samples of all things should be preserved in the school museum.

> वर्णलिंगवयोऽवस्था: परिक्ष्यंच बलबलम् ।
> यथास्थानं यथायोग्यं संस्कारान् कारयेत् सुधी: ॥

Before using any materials, they should be properly improved after taking into consideration its color, gender, age, condition, strength / weakness, place of origin and suitability. Meanings of these terms are given below:-

- Color (Varna) is the class to which a thing belongs by its natural properties.
- Gender (Ling) is a mark denoting a peculiarity in quality due to the formation of the body.
- Age (Vaya) is the effect produced upon a thing by the action of time.
- Stage (Awastha) is the condition ~~reached~~ attained ~~owing~~ due to improvements it has undergone.
- Strength / weakness (Balabal) is the particular capacity or defect of particular thing. Thus tempered steel is very hard but too brittle etc.
- Place (sthaan) is the situation in which a thing is to be used. The same thing when used in different situation requires different qualities as it is acted upon by different forces.

Ability (Yogyata) is the particular quality required to suit the requirements of a particular situation.

All these conditions have to be carefully weighed before the material is prepared for use in a structure.

(2)Design

In designing engineering works the great peculiarity of Indian methods is the consideration not only of physical comfort but also of material case. What effect would a thing produce on the mind of the beholder and the occupant is a matter the Indian engineer has to consider. In designing a house an Indian engineer has to consider whether the structure would last till the owner is again in an affluent condition. Indian houses were designed to last at least 200 years and palaces and such other important structures for at least 600 years. The dimensions of posts and beams are fixed according to this requirement.

'Safety first' is the motto of Indian engineers. Houses having large quadrangles with wall are safest. Thick walls all round full light, air, water etc from the open space of the quadrangle , all these go to make an ideal Indian house. Indian roads are divided into five parts viz.

- First is in the centre for elephants.
- Second and third on both sides for horses.
- Fourth and fifth on the outskirts for foot passengers foot paths .

These are rules laid for regulation of traffic at the junction of roads. The main trunk roads are to pass north and south so that shade can be had by

oot passengers all day long except noon when anybody goes out very
arely.

n designing flight of steps, a broader step is introduced after every four or
ive steps in order to give relief to knee joint. Such steps are at least twice
road as the others. In all the flights of steps on the bathing ghats on river
s also those leading to mountain- tops, this practice is invariably
ollowed.

emples are designed to meet these requirements:-

- The icon (Moortee) is intended to satify the sense of touch.
- The inner door (~~Garbhadwar~~/Garbhagruha) is intended to satisfy the sense of taste. Tirtha Prasad is distributed in this place.
- The inner canopy(Antarmandal) is intended to satisfy the sense of smell as sacrificial tree are blazed here.
- Hara os intended for the satisfaction of the sense of sight. The idol is seen from this place. The ideal of Indian God is not the idol but the lamp that burns near to it. As the lamp devotes its whole existence to the distribution of light so a man is to devote the whole of his life in doing good to the world.
- Padasootra is open space for going round the temple.
- Sabha is the place where prayers are heard or ~~sung~~ chanted.
- Mandap is the place where the ideal devotee is installed as a model for all to copy.
I. Bahirmandal is the place where the poor are fed and cared for.

n designing a structure, Indian engineers specify that the most important
arts should be made very strong and the remaining parts so as to suit the
urse of the owner. No revised estimates are allowed by Indian authors
who specify that all the requirements of the owner must be satisfied within
he amount he intends to spend. The engineer is at liberty to select the
naterials suited to purse but he has to satisfy all comforts and
equirements. A king requires as much food and warmth as a poor man
loes but their financial capacities differ considerably. The engineer
herefore fulfills the requirements and comforts of both by using different
naterials suited to different persons.

t is not easily possible to get all good qualities at once. But the artisan has
o avoid as many defects and secure as many advantages as possible.

अल्पदोषं बहुगुणं कार्यं कर्म

प्रयत्नत: ।

(3) Execution

(a) Selection of a site- the site for a building or residence is selected on the following considerations:-

- The kind of land and its neighborhood.
- The vicinity of water and its supply.
- Kinds of plants surrounding the site.
- Kinds of beasts and birds nearby.
- Kinds of men in the vicinity.
- Gradient and the directions of the prevailing winds and flow of water.

(b) Testing of foundation- The foundation of a building is tested in the following ways;

- By the absorption of water.
- By the compaction of the soil.
- By the direction of currents of air.
- By the absorption of heat and light by the soil.

(c) Selection of materials- The materials are selected , as already noted, by testing their color. It is then determined what material is useful in the particular place. It is then wrought to suit the requirements and this process is known as Sanskar. There are 16 important Sanskars and 24 secondary sanskaras to the first class material. The least sansakaras are 4 and even the worst material undergoes these 4. The strength of a structure is that od its that of its weakest part and hence Indian engineers specify that the whole structure should be of uniform strength.

(d) Plinths- These are required to make up levels, to add to the security against floods etc, and to enhance the beauty. Various designs for plinths are recommended and to suit these, different chairs, posts , arches, lintels doors etc are specified so as to have a uniform impression from the bottom to the top.

(e) Doors and windows- The dimensions of doors and windows are fixed with reference to the number of storeys, the height of one storey and the width of rooms. A uniform light is desired and with this object the heights of the lower storeys and therefore the dimensions of the doors and windows are kept greater than those of the upper storeys.

(f) Roofs- Different types of roofs are recommended for different climatic conditions. The slope of roof varies from 1 in 12 in Sind to 1 in 1 in Kashmir to suit the rainfall and wind conditions of the various Indian districts.

(g) Ornamentation- It is very difficult to keep delicate carvings free from dust and Indian engineers, therefore specify that carvings should be used only on structures of well- to- do persons where cleanliness can be easily secured. Carvings are to be used in temples, palaces and not in ordinary houses.

In conclusion it may be noted that Indian students ought to study Indian texts. Indian texts prescribe Indian materials, indigenous methods and are suited to local manners, customs and conditions. When the people are able to satisfy most of their wants by indigenous products, then and then alone can be prosperous. Indian texts specify Indian wood, Indian stone, Indian metals and take into consideration Indian conditions. Indian manners, habits, customs and religious practices are the outcome of these conditions and therefore Indian texts are worth to study by Indian students. New ideas new inventions and discoveries are also to be studied and used but only in addition to Indian ones. In the present system of teaching the Western element is preposterous. Indian engineering graduates of the present day know how to build a bungalow, design
a coffin or decorate a church but are quite ignorant to building of Indian houses designs, tomb of Sannyasis or construction of temples. They know the properties of oak and Pine trees but not of those of Indian trees. Indian building stones, Indian soils, Indian colors, deserve to be tested by modern methods as also the prescriptions of ancient authors to be verified.

The great advantage of study of Indian Engineering is that it embodies the experience of so many ages. Terraces constructed on the present methods always leak but terraces constructed according to Indian specifications do not. Tempering of tools according to Indian specifications is more lasting than that according to the western specifications. It is true that Indian authors use minerals very sparingly and amongst metals they depreciate iron as the worst and easily perishable metal. The experience of animal and vegetable materials as recorded by Indian authors is vast.
Indian authors advise that engineers ought to be able to treat ordinary accidents on their works and for this purpose devote chapter on treatment

of wounds bruises and diseases(VraNa vyadhinirakaraN). Many of these remedies are very simple and effective.

They require ordinary plants and herbs that are available everywhere and can be had anywhere in India.

Chapter V

Mechanics of Five Elements

Aryan Physics analyses root elements into five basic categories ie Akasha, Vayu, Tej , Aap & Pruthwi . In these five categories again 147 elements are classified . Present known 117 elements for modern science come under Aal tatwa category . These five elements have their own properties based on that mechanics of these five elements is designed . Space,Air,Fire,water & earth are most outer representations of these five root elements . Though they are outermost , they carry same properties like the root elements . Here in this chapter we will see how ancient Aryans used thsese five elements & subjugated them to acchieve work

Fig.1

स्वयंवाहकमेकं स्यादुमकुत्प्रेयं तथा परं ॥

अन्यदंतरितं वायं वात्यमन्यद्दूरतः ॥

समरंगण अ० ३१

Vayu is mobile by itself and is the conveyance of all motion, Tej requires to be moved only once and then it goes on moving to the end of world. Aap can be moved in any direction by giving an impetus in one direction and Pruthwi has to be mo ved by close contact only, it would not now move otherwise. Aakash is as stated above nivrutta (horizon)repungent to motion or has no room for any change. Aakash with motion is Vayu which is the very essence of motion.

Fig 2

गुरुत्वं पृथ्व्यांजलवृत्ति । न्यायसूत्राणि, तर्कसंग्रह

Wherever there is motionVayu exists and
there can be no motion without Vayu. Heat and light is Tej . Heat and light
mean motion of a specific degree. Vayu when it comes to the stage of
motion it --- Thus Aakash , Vayu and tej --- anologus to one another.—but
in a more tan--- not what we call --- Vayu mixed with particles of Aap & --
no water, just like Tej in creation and of Pruthwi and AAp in it is just as
that blue colour of sky is due to the reflection of light on those
intimate particles

Fig.3

प्रकृत्या पार्थिवं स्थिरं शेषेषु सहजा गति: ॥
अत: प्रायेण सा जन्या क्षितावेवप्रयत्नत: ॥
समरंगण अ० ३१

Vayu has neither colour nor weight and what we sense as colour or weight
is due to the minute particles of Prutwi and aap that are in suspension in
what we call atmosphere round the earth and close to it..

Fig.4

दंडश्चक्रैश्च दंतैश्च सरणिभ्रमणादिभि: ॥
शक्तरुत्पादनं किं वा चालनं यंत्रमुच्यते ॥ यंत्रार्णव

Pruthwi is naturally inert and has no tendency to motion . All Machanics
therefore means creation of motion in Pruthwi by artificial means adopted
for producing this motion are called mechanical contrivances or Yantra.

Yantra is contrivance consisting of Danda Lever, chakra or pulley,Danta
toothed wheel, Sarani- inclined plane and bhraman- screw and is required

or production of Shakti-(power or motion) or changing the direction. These mechanical contrivances also have asequence suited to the five elements as under,

1. Danda Lever is specially useful for Uchhatan or breaking or stirring of the Pruthwi.

2. Chakra or pulley is specially designed for Vashikaran controlling the motion of Aap

3. Danta- toothed wheel is particularly useful for sthambhan-stoping of Vayu.

4. Sarani- Inclined plane for Jaran- bringing toghether of tej

5. Bhraman –Screwis mainly useful for maran- killing or annihilation of Aakash time and space.

Thus the two principal elements have the five mechanical contravances for producing particular kind of motion The orientation or properties of these elements require a particular sort of machinery and is capable of a particular sort of motiononly. For example rays of Tej can only be converged or diffused & no other kind of motion is possible in them and this can be done by means of inclined plane only A mechanical contrivance Yantra consists of three parts,viz i power or producer of motion, ii fulcrum or the pin joining power and work iii shakti the power doing the work The power by means of which the work is done is generally less in quantity or more suitably available than the work beejak is the point or axis at which beej and shakti meet each other and shakti is the ability to achieve the desired end or the moment to moment, is only a chakra having projecting teeth to afford better hold for the power. In a chakra only frictional part of power is available for work and some times this is not sufficient to produce the necessary motion and slipping occurs. The teeth are intended forany such contingency.. Teeth afford a better hold and a more accurate attachment than friction. Sarni is only an angular lever nwhich insted of shifting of lever beej and shakti slide along with it. In this case the Aksha is fixed and not movable as in chakra and this is the point or rather a line at which the two planes formed by the two arms of the lever meet. Bhram is nothing more than a XXX wound round an axis like a creeper winding round a tree. as all the five elemnets are only different isometric states of Aakash all mechanical contrivances Yantra are different manipulations of Danda to suit the particular circumstances of the constitutions of the elements forming universe.

The various objects for which machines are designed are enumerated as under,

1. In some machines some particular action has has to be achieved continuously for any length of time.work actually executed .Danda lever requires the kilak to be continuously shifted in order to produce continuous work and in order to avoid this frequent changing chakra comes in use, In Chakra the Aksha axis is the focus of all the points at which kilak is required to be moved from.

2. In some machines a particular action has to be done at a particular moment at regular interwals or in a fixed sequence with certain other actions..

3. In some, a particular sound has to be produced or magnified in certain ratio or way.

4. In some machines the form or feeling has to be similarly developed or magnified to suit particular effects.

5. In short various kinds of motion have to be produced for attaining various objects and as the objects tto be attained are inumrable the motions to be produced are also innumarable. The various kinds of motion may however can be classified as

Fig.5

कस्यचित्का क्रिया साध्या, काल: साध्यस्तु कस्यचिद् ॥
शब्द: कस्यापि चोच्छ्रायो, रूपस्पर्शौ च कस्यचिद् ॥
क्रियास्तु कार्यस्य वशादनंता: परिकीर्तिता: ।
तिर्यगूर्ध्वमधः पृष्ठ पुरत: पार्श्वयोरपि ॥
गमनं सरणं पात: इति भेदा: क्रियांद्भवा: ॥

समरांगण अ॰ ३१

•Tiryak or slant
• Urdwa or upward
• Adho – downwards
• Prushtha- Bacward
• Purat- Forward
• Parshwatha – Sideways
These six motions are again classified in three as
1. Gaman –motions backward or forward
2. Saran- Motion left or right sideways
3. Patan –Motion up or down/
A machine must be able to to produce any of these motions at any desired moment. A machine is considered better as it is more able to give satisfaction in these acts .

The chief qualities which a machine should posses are devided as under,

Fig.6

यथावद्बीजसंयोग:, सौश्लिष्यं, श्लक्ष्णतापि च ॥
अलक्षता, निर्वहणं, लघुत्वं, शब्दहीनता ॥
शब्दे साध्ये तदाधिक्यं, अशौथिल्यं, अगाढता ॥
वहनीपु स‍त‍षु सौश्लिष्यं, चास्पलद्वृति: ॥
यथाभिष्टार्थकारित्वं, लयतालानुगामिता ।
इष्टकालेर्थदर्शित्वं पुन: सम्यक्‍व संवृति: ॥

अनुल्बणत्वं, तादूप्यं, दाढर्यं अस्रणता तथा ॥
चिरकालसहत्वं च यंत्रस्यैते महागुणा: ॥

समरंगण अ० ३१

1. Application for force suited to the work and time
2. Proper contact
3. Smoothness

4. Requiring no attention
5. Continuous action
6. Lighness or requirement of as little force as possible
7. Silence
8. Loud sound when that is the object to be gained
9. No looseness
10. No sticking fast or clogging
11. Proper attachments of all the parts particularly bolts used for transmitting motion
12. No intermitence or break in action
13. Perfect attachment of the desired object
14. Adjustment as to time in simultaneous actions
15. Doing desired action exactly at the desired moment
16. Return to the normal condition at other times
17. No peeling off
18. proper form and shape
19. Strength
20. Softness and elasticity
21. Long life

These are the twentyone chief qualities which every machine must possess. The more the qualities a machine possess the better it is .In designing and constructing machines the objects should alwaysbe kept in view, In a mill (mahayantra) or factory many things have to be done simultaneously and the production of one thing is interdependent on the other, the various parts have ,therefore, to be designed to keep pace with one another for any length of time continuously.

Fig.7

यदृच्छया प्रवृत्तानि भूतानि स्वेनवर्त्मना ॥
नियम्यान्यस्मिन्नयति तद्यंत्रमितिकीर्तितं ॥
तस्य बीजं चतुर्धास्यात् क्षितिरापो नलोनिल: ॥
आश्रयत्वेन चेतेषां त्रियदप्युपयुज्यते ॥

समरंगण अ० ३१

A machine is a contrivance that forces the elements to quit their natural courses and adopt that which would compel them to do the work on their way. IIn a machine the chief ingradient used are Pruthwi, Aap, Tej and Vayu while Aakash being pervading is naturally used as the basis of all. AAp, Tej and Vayu are the principle motive agents and Pruthwi is the element which is moved for the particular object that is to be attained. The elements are used in constructing machines as noted below:-

Fig.8

कुब्यंकरणसूत्राणि भारगोलकपीडनं ॥
लंबनं लंबकारं च चक्राणि विविधानि च ॥
काष्ठं च चर्मवस्त्रादि पृथिवीकर्मचोच्यते ॥

समरंगण अ० ३१

Pruthwi is used in machines for the construction of envelopes, for ropes, for weight, for use as a brake (pressing against a ball) for leghening, for transmission , for wheels of various sorts, for wood, for skin and for clothing etc. Pruthwi is to be used for all purposes in which direct motion is not produced but in which other materials are kept or contained and motion is transmitted from one thing to another: it is also used for stopping motion. In short Pruthwi is an inert body and is Dabhya or obidient to the commnds of the employer. It has no will of itself except Gurutwa, the all prevading or universal force which is the prime mover or first cause of the whole universe, This gurutwa is omnipotence itself as it is the power which apperaed first in creation and which converted the Nirgun into Saguna ,the intangible into tangible things.

Fig.9

धारा च जलभारश्च पयसोभ्रमणं तथा ॥
यथोच्छ्रायो यथाधिक्यं यथानारंध्रतापिच ॥
एवमादीनि कार्याणि जलजानि प्रचक्षते ॥

समरंगण अ. ३१

The flowing force of water, it;s weight and the rotating power are qualities which are chiefly used in Hydraulic machines, When water flows with a force it conveys things with it or it pushes things when water drops from height it exerts an impact and when it is pushed in one direction it reaches in all directions and these are the properties of water that are used in water machines. When a thing floats in water it means that water lifts it up with a pressure equal to its own weight.. When turbines and sprouts rotate it means water reacts on the pipes while getting out at the open ends

Fig.10

संग्रहीतं च दत्तं च पूरितं प्रतिनोदितं ॥
जलं बीजत्वमायाति यंत्रेषु जलजन्मसु ॥

समरंगण अ. ३१

Water can be stored , it can be supplied, it can be directed or compelled to react and in this way its force can be used as power in Hydraulic machines. This storage etc. has to be resorted to because the power of water is greater as it is sutuated at a higher levelor as it is available in a large quantity or as it is conveyed in a more water –tight tube.For any kind of flow these conditions which have to be fulfilled. Even in education which is a flow of knowledge from the preceptor to the pupil these conditions are required to be fulfilled. The preceptor is to be Shroutriya one having a large knowledge, Bramhanisht , one having knowledge at great height of practiced experience and the pupil is rwquired to be Bramhchari, one having all the from the body tightly closed and then only the flow is properly effective and strong.

Fig.11

प्रेरणंचाभिघातश्च विवर्तो भ्रमणं तथा ॥
अत्यंतमूर्ध्वगामित्वं मारुतोत्थेषु संमतं ॥

समरंगण अ. ३१

Pushing,, Striking, Rotation, Blowing and Rising to a very great height are
the qualities of Vayu elements that are used in machine in construction .
The Vayu is enclosed in pipes and compelled to push things on its way, it
is made to strike on vanes or blades and move them, it is forced to produce
rotary motion by reaction or it is allowed to carry things while blowing in
a particular direction. Some times it is used to lift substances to
great heights. In short in practice Yayu is a more subtle . Aap and the
qualities and laws of both are the same

Fig.12

प्रेरितः संग्रहीतश्च जनितश्च समीरणः ॥
आत्मनो बीजतां याति एवमन्यत्प्रकल्पयेद् ॥
यंत्राणॅव

Vayu can be directed, it can be stored or it may be created or produced and
used as power. When it is conveyed in pipes it can be directed to flow in a
particular place as in the blowing machine. It can be stored and made to do
a particular work on its way as in a a steam engine or hot air apparatus. It
may be produced at the place or its use as in a blasting operation. These
are the ways in which Vayu can be used and necessary arrangements to suit
a particular case have to be made by an Engineer.

Fig.13

ताप उत्तजनं क्षाभः शाभश्च जल संगजः ॥
एवमाद्याग्नि बीजानि भवंति......॥ यंत्राणॅव
प्रत्येषकंच जनकं प्रेरकं ग्राहकं तथा ॥
संग्राहकं च यंत्रेषु बीजं स्यादनलोद्भवं ॥
समरंगण अ ३१

Heating, exploding, expanding and evaporating Aap are the qualities that can be had from heating property of Tej in mechanical apparatus. Things may be heated and made soft or pliable as in joining different metal pieces , the gases may be made to expand and do work by their force , blasting materials may be ignited and made to explode or ignite things may be converted into gaseous state amd made to work . Heat of Tej can be used in any of these ways. The lighting power of Tej is used as an experiment or as a generator or as a director or as a holder or as astore in Tej machines , Tej is used not only as heat but also as light. The power of heating is used in machines as noted above and we here have to see how lights can be used as a power, light is used for lighting things or exiting sight to percieve things invisible as ib a dark room or tunnel, it can be used at an active agent in changing colours of things, it may be used as director as in a reflector or mirror,it may be used as a holder of images as in a telescope or microscopeor it may be used as a store as in a burning glass. By means of proper manipulation of light , invisible things may be made visible, small things may be magnified, distant objects may be brought closer, special signals may be flashed ,or inflamable materials may be set on fire. Thus Tej is used in two ways and these differ only in their subtleness, light being more subtle than heat.. From this heating property of Tej. we come to the properties electricity. Mitravarunou, when a thing is unequally heated that is when one of the parts is hotter than the other an electric current is set in motion or electric equilibrium is disturbed. Equilibrium is a state of calmness and may mathematically be expressed or represented by zero. When this equilibrium is disturbed two equal and opposite forces are generated and these are termed as Ghan (Mitra) or the positive, Rhun(Varun) the negative repspectively.. The whole universe is evolved from equilibrium and is therfire composed of two parts ,Dand and XXXX , thus Mitra and VArun pervade the whole of the world. Mitra is seen in Gati, Tejha, UChha, Jal, Shassan,Raktadrav, Anrssar, Punsar and /,Dharm, things. & while Varun is seen Shtiti, Tamha, Neech, Pruthwi, Anshan,Hariddrav, Bahissar, Stree, Asur & Adharma things and thus the whole world is divisible into two two complementary sections,. Mitra and VArun are defined in the Vedas as under,

Fig.14

Mitra is (Poot) pure and Careful (Daksha) or intelligent, Varun is eater or consumer of bad things. The earth is hot in the interior and cold on the outside. It is hot on the the part towards Sun and cold on the part away from it. It is hot at equator and cold at the poles. On this account the eart is surrounded by an electric current and it is a magnet,
It's hotter part is Mitra and coooler part Varun. The north pole is Mitra and south pole Varun. The power of this force was investigated by two ancient Rhushis (sages)viz, Vashisht and Agasti and hence they are called Maitravarni or Electrical engineers.Electricity and Magnetism have very close connection with each other and, they are very easily convertible one into other, By the passing of this electric force through many substances they are decomposed into their component parts and liike things are attracted to the like poles. Mitraalways collects together the higher things and Varun the lower ones in the scale. It was Angarup who first decomposed parts of Oxygen(Pran) and Hydrogen (Udan) gases.. He was inventor of electroplating process and for this purpose he used hundress of copper and zinc battery which is described in Agastya samhita as under:-

Fig.15

संस्थाप्य मृण्मये पात्रे ताम्रपत्रं सुसंस्कृतम् ॥
छादयेच्छिखिग्रीवेन चार्द्राभिः काष्ठपांसुभिः ॥
दस्तालोष्टोनिधातव्यः पारदाच्छादिदस्ततः ॥
संयोगाज्जायते तेजो मित्रावरुणसंज्ञितं ॥
अनेन जलभंगोस्ति प्राणो दानेषु वायुषु ॥
एवं शतानां कुंभानां संयोगःकार्यकृत्स्मृतः ॥
अगस्त्य संहिता

The word Pruthwi in connection with Mitravarunou means an earthware vessel used for storing Ghrut – ghee and Apsara means a pot impervious to water. A battery cell is called Kumbha and Agrup is called Kumbhodbhav (a pot produced by a person or a person encircled by kumbha the batteery cells) Indian scientists distinguish the following sorts of electricity:-
1. Electricity produced by flapping of skins or silk- Tadid
2. Electricity produced by rubbing jewels or glass- Soudamini
3. Electricity produced in clouds or vwater vapour- Vidyut
4. Electricity produced by hundreds of cells of any batteryor pieces of metals etc=
Shatakoti or Shatakumbhi
5. Electricity stored in a storage cell – Hrudini
6. Electricity produced by magnetic iron bars= Ashani
Human body is understod to be a magnet by circulation of electric currents round it and the principle of saluting a preceptor by touching his feet with the head and holding the right foot with the right hand and left foot by left hand by a pupil is based on the principle of joining two magnets in the most effective manner as detailed below

There are three possible ways of joining two magnets as noted below:-

1. Joining two north poles and joining two south poles in tandem. In this there is some repultion in the two similer poles and the product is consequently little less . This is the case when two human beings embrss each other,
2. Joining the north pole of the one with the south pole of the other and vice versa. The two opposite poles cancel each other and leave no power in the combination. The combination has no magnetic strength at all. This is the case when two persons salute prostrate at each other;s feet.
3. Joining two magnets in a series. Here the north pole of one is joied to the soutth pole of the otherand the two magnets togrther form one long magnet. Here the strength is little more than the sum of two indivisual magnets. This is the case in two salutation of the preceptor by his pupil.
In this way Indian scientists utilise the properties of all the different sorts of electricity in their every day life, viz. tadid in aasan or seats, soudamini in ornaments and garlands, Shatakumbhi in electroplating, Vidyut in mesmerising by the sprinkling of water and so on...

References

Samarangana Sutradhara – By King Bhoja
Yantrarnava
Agastya Samhita
Tarka Samgraha

Chapter VI

Automation & Robotics from 10th Century King Bhoja

King Bhoja of Paramara Dynasty from central india was a polymath . He wrote many books on different subjects ranging from technology to yoga and from erotica to astrology. Here we will take into consideration his two works one a poetic composition and the other a technical treatise. The former, which is yet unpublished, is the prose work of fiction, SringSramaHjaH.38 In it, Bhoja himself is the hero and, as it is not proper for a noble soul to indulge in self-glorification, the device is adopted of making a mechanical figure {yantra-putrika), discharge the function. Confirming almost all the details set forth on the one side by Somadeva in his Campa and on the other by Bhoja himself in his book on architecture, this work gives an elaborate description of a yantra-dhara-griha, a fountain pavilion with manifold mechanical works: there are figures in constant action; there is a jalayantra-putrika which scatters a fine spray; dharas which fall like slender lotus-stalks, and in a curve like a bow; mechanical drum-players filling the place with their rhythms; an artificial lotus pond; toy bees which keep humming; yantra-vriksha or trees, with monkey figures; a pond filled with replicas of cranes bending over and getting deceived by fishes which come near and move on; artificial tortoises diving and coming up every now and then; and yantra-orchestras.
The Samaranganasutradhara ascribed to Bhoja is, in many ways, a rare treatise in Sanskrit literature; besides the Arthasastra, it is the only theoretical text that has substantial information on our subject; its value, however, is greater than that of the Arthasastra, as Bhoja goes into the details of the construction of these yantras and explains at the beginning the principles under lying, yantras.
Chapter 31 of the Samaranganasutradhara is called Yantravidhana and its 224 verses are wholly devoted to a description of various mechanical constructions. Bhoja opens with the definition of yantra, that it is so called because it controls and directs, according to a plan, the motions of things that act each according to its own nature. As we have seen and as Bhoja explains, it is from the principle of control, yam, that the name yantra is derived

The next topic Bhoja deals with is Bija. Bija means a constituent element. The constituent elements of a yantra are four: Earth, Water, Fire and Air, Ether being the basis and medium of action,
Bhoja then discusses whether Suta or mercury, which is an indispensable ingredient, is to be held as one of the Bijas along with Earth, Water, etc. Some earlier writers had counted it separately as a Bija. Bhoja says that mercury is essentially Parthiva, i.e., to be brought under Earth, though one might find it in the liquid state and also possessing a property causing motion, like wind. (6-8 verses).
The machines are then divided into two classes, Svayam-vahaka, automatic and Sakrit-prerya, requiring occasional propelling; most machines combine these two features (13). Another classification is: the concealed, Antarita or Alakshya, i.e., the principle of its action and its motor-mechanism are hidden from public view; the Vahya, or the machine to be carried by another; and the third, which is really obscure, but may be interpreted as the distant or proximate, meaning thereby the place from which the machine acts (10-11). Some move many persons and things while others require many persons to move them (49). That machine is best whose principle of action is concealed, which achieves manifold purposes and which excites wonder (12).

After a lavish encomium on the comfort and advantages to be enjoyed through the machines, Bhoja proceeds to describe the constituent elements called Bijas of each variety, the Parthiva, Taijas, etc.; while all the elements may be used for a single yantra, it is to be named after the dominant constituent (21-25, 42). Then are mentioned the materials: metals—tin, iron, copper and silver—and wood, hide and textiles; the parts and the principles: the wheels and the rotation; the suspenders and the hangings; the rods, the shafts and the caps; the tools; and the work: measuring, cutting etc.—all these are also to be included under the Bijas o a Parthiva-yantra (25-27). The application of fire-bijas on earth-machines comprises heating and boiling; of water, mixing and dissolving, pouring of and filling with water, and providing a belt of water. Height, size, closeness and motion towards a higher plane are spatial featured in Parthiva-yantras. The element of air is to be applied through bellows, fans, flaps, etc. (28-32). Similarly in machines which are mainly jala-yantras the use of timber, hide and metal forms the Parthiva element and so on. (33-41 As, however, machines have to take some shape and possess a body, the Parthiva is an important constitutent (43-44).

The merits of a good machine, yantra-gunas, are as follows;—37

1. Proper, proportionate utilization of the elements constituting it.
2. Well-knit construction.
3. Fineness of appearance.
4. Inscrutability.
5. Functional efficiency.
6. Lightness.
7. Freedom from noise where it is not part of the scheme.
8. A loud noise when noise is intended as an end.
9. Freedom from looseness.
10. Freedom from stiffness.
11. Smooth and unhampered motion.
12. Production of the intended effects
curios)
13. The securing of the rhythmic quality in motion
14. Going into action when required.
15. Resumption of the still state when not required (chiefly in cates of ! the pieces for pastime).
16. Freedom from an uncouth appearance.
17. Verisimilitude (in the case of bodies intended to represent birds, animals etc.).
18. Firmness.
19. Softness.
20. Durability. (45-49).

Now to the Karma or action of these machines : Machines are characterized sot only by «n action peculiar to each but also by the particular times when they art to operate. The specialty of some is sound; of some, height, form or touch; and so on. Action is across, upward, downward, backward, forward, on either side, speeding and crawling. Another factor is the time taken for the action. In sound, the factors are variety, the quality of pleasing or the capacity to terrify. In contrivances for pure entertainment, music, dance, drama and imitation of different things and beings are the main factors, each, like music and dance, having its sub-varieties. In motion, going up and coming down. For the reproduction of whole themes in machinery, Bhoja instances the fight between the Devas and the Asuras, the churning of the ocean, Nrisimha killing Hiranyakasipu, races, elephant-fights, a mock-army, etc.- (.50-62).

Among fittings for utility, beauty and sport, various types of shower-fountains (dhard-grihas), swings, pleasure-chambers, mechanical carriers and servants, balls and magic (?) are mentioned. (63-64).

Bhoja now proceeds to describe some of the things that can be accomplished through yantras:

1. Five storeys could be arranged and the bed placed on the ground floor made to go up to each higher floor at the end of each watch of the night. (65).

2. Another pleasure contrivance is the couch called Kshirabdhisayana, in which the serpent-like bed goes up and down by the soft action of air, like that of the serpent's breathing. (68-69).

As examples of miracles that could be worked through yantras, Bhoja mentions the production of fire m the midst of water and vice versa; effecting the complete disappearance of a thing present before one and the projection before one of the view of a thing not present before him (67-68). How these were done is not stated.

In Verses 66-67, a kind of chronometer is described; there is a circular device in which, in a broad open vessel, there are thirty, probably ivory figures, or tooth-like pieces lying flat all along the circumference; the whole thing is revolving; in the centre is the figure of a lady, who wakes up one figure or piece for every Nadika.

Another chronometer-like object is described in Verses 70-71: there is a rider on a chariot, an elephant or any other animal; for a fixed time, say, a Nadika, the rider on his

mount goes round and at the end of the Nadika, the chronometer strikes. An astronomical model called Gola is then described (69-70}, in which there are needles and the day and night movement of planets is shown.
18
Among mechanical contrivances given is a lamp into which, at set intervals, a mechanical figure goes on pouring oil. An additional feature of this is that the figure keeps on circumambulating to a definite musical rhythm (71-72).

Other entertaining yantras are speaking, singing and dancing birds, a dancing elephant, horse or monkey, water going up and descending, a mock-fight and others worked by the manipulation of air. (73-78). Bhoja concludes this section with the observation that not only these, but many more similar contrivances could be invented; even movements impossible in actual life are possible in yantras. Regarding the actual process of making these, Bhoja says that, though it is not set forth, the constituent elements and basic principles have been mentioned, so that men of imagination could easily construct these. Silence on the actual mode of construction is said to be for preserving this important knowledge, for giving a material advantage to the architects, and for enhancing curiosity about these yantras. (79-81).38

Another interesting statement that Bhoja makes is that some of the yantras described by him are those actually seen by him (drishtani); more

important is his information that follows, viz., that he would now proceed to describe some more, handed down from earlier masters. (82-83):—
After speaking again of the Bijas or constituent elements and the wonder and pleasure of these yantras, Bhoja refers to the Sutradharas or chief architects who do these; their qualifications are set forth as: (1) traditionally handed- down knowledge, (2) skill combined with schooling under masters, (3) practice and application, and (4) imagination. (85-87). Five classes of yantras are then mentioned as constituting the five sections of the science of yantras, yantra-6&slra-a4hikdra. (86).; the lines not being free from corruption, we are not able to make out these five classes of yantras to which reference is again made later; but movement, such as rotary; material, such as wood; purpose, such as the exhibition of dexterity and the satisfaction of curiosity; utilitarian value and pleasure, as in swings; and form, as in the round yantras (cakra) is about the best meaning that I can extract out of the context here.39
A series of yantras now follows with some details of their manufacture:40
1. A wooden bird in whose hollow body is placed a copper contrivance one inch long and one-quarter inch high, of slender cylindrical shape, in two well-joined halves allowing a hole at the centre along which air passes when the bird moves, creating a pleasing sound (89-90),
2. The next is actually noted as a bedroom accessory. In the hollow of the bird above mentioned is placed a small drum-like piece in halves and with an air-passage as in the previous yantra; the interior device is to be loosely hung and as the bird oscillates, a highly pleasing sound is created which reduces the anger of the ladies who are cross,. (91-92)
Other bedroom accessories are various mechanical musical instruments which sound automatically on the principle of stopping and releasing air according to plan. (93-94}.
The third class of yantra described is the aerial vehicle which runs on mercury as fuel .
The fourth category comprises male and female figures designed for various kinds of automatic service. Each part of these figures is made and fitted separately, with holes and pins, so that thighs, eyes, neck, hand, wrist, forearm and fingers can act according to the need. The material used is mainly wood, but a leather cover is given to complete the impression of a human being. The movements are managed by the system of holes, pins and strings attached to rods controlling each limb. Looking into a mirror,

playing a lute and stretching out the hand to touch, give pan, sprinkle water and make obeisance (101-4) are the acts done by these figures. It is one such that provides the mechanical fan in the Yasastilaka Campu. Similar robots are used for the palace guard; one such stands at the gate with a baton, sword, iron rod, spear or other weapon and prevents the entry of outsiders. This can quickly and quietly kill thieves who break into the palace at night.

Bhoja closes this section with a reference to military equipment in forts, bows, *sataghnis* and a weapon newly mentioned by Bhoja, the *Ushtra-griva* (camel's neck), resembling probably the modern cranes. He also indicates here the classification into *Guptyartha* and *Kridartha,* protective military yantras, and yantras for sport and entertainment respectively . (108).

The sixth series now taken up by Bhoja in the *Kridatha* class, is the fountain, *vari-yantra*, described in some detail by Somadeva, and dealt with by Bhoja not only here, but also earlier under palace architecture. Movement in. the *vari-yantra.* is fourfold: (a) a downward flow from an overhead tank for which a *Pata-yantra* or waterfall-machine, is to be used ; (b) *Samanadika* is for the release of water at a higher level from tanks placed at that level; (c) *Patasama-ucchraya* is a contrivance using bored columns for letting down water from a height, and then taking it up through columns placed aslant; and (d) the last, *Ucchraya*, in which water from a well or in a canal on the ground is sent up by u device. (110-14). An artificial object which was probably common in Bhoja's time is a wooden elephant which Bhoja describes twice; earlier he cited it as an example of various wonderful effects that could be achieved through yantras (72-3). It occurs here again under yantras based on the principle of sending water upward, *Ucchraya* (115). This wooden-elephant drinks water placed in a vessel, any amount of it, and neither the intake nor the

water taken in is perceivable. On the *Samaucchraya* principle of circulation of water on the same level is based the underground conduit which brings water to a tank from a distant source (116). Water conduits ir general are described earlier-also in Chapter 18 in connection with the city and residence.

In the class of up-and-down play of water, Bhoja expatiates upon the *Dharagriha*, shower-bower, in the garden ; its popularity hag already been noted, and Bhoja has already given a description of it in Chapter 18; Its great vogue can also be seen from its different types known by distinct names mentioned by Bhoja:

(1) *Pravarshana*, the shower, (2) *Pranala*, the pipe, (3) *Jalamagna*, the sub aquatic, and (4) the *Nandyavarta*, in a special design. These are constructed only in palaces for the King's pleasure. (117-18).

Regarding their construction Bhoja says: (1) They are to be in the proximity of big reservoirs; (2) They should occupy a site with good scenic possibilities; (3) Pipes have to be prepared to double and treble the height and other requirements of the fountains; the pipes should be able to carry water, be free from poise, and smooth inside (119-20). Naturally architectural erections add to the excellence of the fountain-park and parts of the structure are themselves used for the different water-works.

Fine and fragrant timber, Devadaru, Sandal, Sal, are to be used for the woodwork, carved pillars, platforms, projections, windows, cornices, etc. The main items are female figures and models of birds, animals like monkeys, manifold forms with gaping mouths, semi-divine and half-human and half-animal forms, *Nagas, Kinnaras* etc., dancing peacocks, *Kalpavrikshas*, creepers and bowers, cuckoos, bees and swans. In the centre of the flowing stream is to be fixed the main pipe, the exterior of it being made into any charming form according to one's liking. To the top of it is fitted and fastened strongly with *vajralepa*, cement, the devices for taking up water, Scattering and throwing it in a variety of ways (133). The pond is to be filled, for effect, with yantras of animals and aquatic beings, *e.g.*, sporting elephants which do even minute actions like closing their eyes when another throws water on the face (134); other specimens we saw in the description in the author's *Sringaramanjari*. Female figures spraying water from eyes, nails, etc., when those parts are touched are described here {136-37) as well as in the earlier chapter (XVIII. 47-50). The King's seat is right in the centre on a fine stone; he sometimes indulges in a bath, sometimes enjoys the play of water from these manifold contrivances, the *Jala-silpas*, sometimes listens to music and watches dancing here, and, particularly in summer, the fountain is a necessity {139-141).

More specific descriptions of the four types of *Dharagriha* now follow: The main specialty of the first, the *Pravarshana*, the shower, is that it pours down water. Strong figures of three, four or seven men should be set up, with curved tubes; the whole mechanism is fitted with water which is poured out in different ways by these figures (142-46). Bhoja calls this shower-house a pseudo-cloud, *anukaranam ekam jalamucam* (148)—(Somadeva Suri's commentator gives it the name *kriirima-megha-mandira*)—a boon in summer and a feast to the dyes. Kalidasa's reference to the yantra-dhara-griha has already been noted but when he says in his *Meghasandesa* I. 61 (*nesyanti tvam sura-yavatayo yantra-dhara-grihatvam*), that the celestial damsels on the Himalayas would scratch the cloud with their bangles and convert it into a yantra-dhara-griha; he seems to know also the name of this type called after the cloud.

The next variety called *Pranala* is two-storeyed with a single pillar or four, eight or sixteen, built like a *Pushpakavimana* with decorative designs. At the centre below is a water-tank with a big lotus, its pericarp fashioned as the seat of the King; around are female figures looking at the lotus; when the overhead tank is filled and closed, water is poured by the figures on the King sitting on the lotus seat.

The third, *Jalamagna*, is a chamber under water, the idea being that of the submarine abode of Varuna or Nagaraja. A square chamber is built at the bottom of a big and deep water-reservoir, the approach to it being through a subterranean passage. A continuous flow of water above keeps the chamber completely cool and the whole reservoir is full of mechanical lotuses, fishes, birds etc. When resting in this chamber alone or in private company, the King can be seen only by select personal friends and urgent visitors of rank like other Princes or Ambassadors. {157-66).

The last type, *Nandyavarta*, has, in mid-tank, a-big flower-like structure; all around the central floral design, in mid-water, are placed low walls in Svastika designs, providing a sufficient screen as well as a passage, the purpose being to permit playing in the water the game of hide-and-seek (167-72).

The fifth main division of the *yantradhikara* was mentioned at the beginning as *Rathadola*; Bhoja now takes it up. *Rathadola* is a swing or a merry-go- round in which people ride in seats, and enjoy the pleasure of wheeling round. That merry-go-rounds were a common sight is seen also in descriptions like the one we find in the poem *Citrabandha Ramayana* of Verikatesakavi (Tanjore Ms. No. 3772, Verse 6) where the courtesans wheeling round in the *Daru-yantra* in the palace courtyard are described by the poet as stars going round Mount Meru. Bhoja devotes as much attention to. it as to the *yantra-dhard-griha;* here too types are known with distinct names, but Bhoja's descriptions here, though detailed, are not as clear as in the case of the fountains. The varieties are called *Vasanta, Madanotsava, Vasantatilaka, Vibhramaka and Tripura,* (174) and each subsequent type is more elaborate and complicated in its mechanism than the previous one.

In the *Vasanta* type the yantra is planted in a dugout 8 cubits square and 4 cubits deep; both metal and woodwork are mentioned at the base of the yantra where the rotation' mechanism is fitted to a platform. A storey is to be raised on twelve posts; on the whole five machines are to be employed for the rotation, wheel acting upon wheel and the whole moving the storey, designed like a lotus and accommodating the whirling riders (175-87).

In the second, the *Madanotsava*, there is no dugout or underground construction; the storey on the main post provides only for four seats and a man standing below operates the machine (188-94).

In the third, the *Vasantatilaka*, two storeys are to be constructed the second one with much decoration; the mechanism is fitted in the first floor and by the action of wheel upon wheel the top floor revolves (195-200).

The fourth, *Vibhramaka*, provides for increased accommodation and variety of motion. At the base here is a solid platform and a square structure with mechanism; over these is a floor with eight seats, and above these another round of seats; spoked wheels link up the whole erection; the specialty here is, each floor has its own different movements, creating, as the name implies, a complex of circular movements (201-8).

The last, *Tripura*, increases the tiers by one, justifying its name of three cities in air, each higher floor being of smaller dimensions; a large number of connecting links, small wheels and steps leading from one tier to the other are mentioned

Chapter VII

Merucry engine aircraft of King Bhoja

The most curious of the yantras described by Bhoja in this chapter is, of course, the one that rises and travels in the air. From the previous notices of this aerial machine only the barest details of its make-up could be gleaned. The only text that gives us some knowledge of its actual construction is this work of Bhoja. Firstly Bhoja mentions the main material of its body as light wood, *laghu - daru*; its shape is that of a huge bird, *mahavihanga*, with a wing on each side. The motive force is then explained: In the bowels of the structure, below, is to be a fire-chamber with mercury placed over a flame. The power generated by the heated mercury, helped by the concurrent action of the wings which are flapped by a rider inside, makes the yantra go up and travel far *{data)* (95-96).
A heavier (*alaghu*) *Daru-vimana* is then described (97-98); it contains, not one as in the previous case, but four pitchers of mercury over iron ovens. The boiling mercury ovens produce a terrific noise which is put to use in battle to scare away elephants; by strengthening
the mercury chambers, the roar could be increased so that by it elephants are thrown completely out of control. This specific military use of aircraft against elephants tempts one to suggest that the *Hasti-yantra* advocated by Kautilya against elephants was something like the heavier *Daru-vimana* described by Bhoja.

लघुदारुमयं महाविहङ्गं दृढसुश्लिष्टतनुं विधाय तस्य।
उदरे रसयन्त्रमादधीत ज्वलनाधारमतोऽस्य चाग्निपूराम्॥
तत्रारूढः पुरुषस्तस्य पक्षद्वन्द्वोच्चालप्रोज्झितेनानिलेन।
सुप्तस्यान्तः पारदस्यास्य शक्त्या चित्रं कुर्वन्नम्बरे याति दूरम्॥
इत्थमेव सुरमन्दिरतुल्यं सञ्चलत्यलघु दारुविमानम्।
आदधीत विधिना चतुरोऽन्तस्तस्य पारदभृतान् दृढकुम्भान्॥
त्रयःकपालाहितमन्दवन्हिप्रतप्तहकुम्भभुवा गुरौन।
व्योन्मो भगीत्याभररात्वमेति सन्तप्तगर्जद्रसराजशक्त्या॥
वृत्तसन्धितमथायसयन्त्रं तद्विधाय रसपूरितमन्तः।
उच्चदेशविनिधापिततत्पं सिंहनादमुरजं निदधाति।
स काऽप्यस्य स्फारः स्फूरति नरसिंहस्य महिमा
पुरस्तादस्यैता मदजलमुचोऽपि द्विपघटाः।
मुहुः श्रुत्वा श्रुत्वा निनदमपि जम्भीरविषमं
पलायन्ते भीतास्त्वरितमवधूयाङ्कुशमपि॥

There may be some lacunae ii\ the description, and Bhoja does not fail to mention that some vital knowledge is kept back as a secret, an idea which we noticed in the *Brihatkatha* story also. It is, however, clear that mercury vapour ought not to be confused as providing any lifting power; it was evidently converted into mechanical power, and the machine must have risen, as is expressly slated here, and implied by the mention of its cock-shape in the *Brihatkatha* story, by the flapping of the wings, and further movement must have been due to the manipulation of the wings and the flow of air itself, on the analogy of the flight of birds.42

An important point to be noted in Bhoja's treatment is that he discusses the views of some earlier writers {XXXI.6), expressly mentions some of his yantras as having been described by the ancients {84), and refers to *yantra- sastra-adhikara* as comprising five sections (88), All this implies the existence of a technical literature on the advanced yantras. Dandin, we noted, mentioned earlier authors on the subject, Brahma, Indra and Parasara . Around three centuries before Bhoja, Gorakshanath describes his mercury propulsion engine in Goraksha Samhita , metallurgical work from 7th centruy .

Chapter VIII

Advanced Aircrafts of King of Mahabalipuram

Sanskrit Poet Dandin(7th century AD), who came soon after another great sanskrit Poet Bana and became famous as court poet of a King of from Kanci in South India, has much more and more varied information to give in the autobiography attached to the opening portion of his prose work, the *Avantisundari*; this forms the full text of the truncated *Dasakumaracarita*, and the introductory portions referred to are still in manuscript. " In one of the introductory verses praising -earlier poets, Dandin introduces yantras; praising the author of the *Mahabharata*, he says that, but for the knowledge that Vyasa infused in us, we would merely be human machines

In Dandin's own life-story, a gifted architect Lalitalaya, son of another eminent architect Mandhata, is introduced and the achievements of these two, which form the subject of the amazed talk of the people, are set forth. It is said here that the father excelled even the Yavanas, from which we have to deduce that Mandhata and his son Lalitalaya were natives of the soil. Once the father, anxious lest his young son might be hungry, rushed to him in an aerial car, evidently from a distance where he had been at work ; which shows how casually the architect took his personal equipment of an aerial vehicle.

The son, who is the actual character figuring in the narrative, and is said to excel his father, is credited with the following achievements the description o/ which forms a brief treatise on yantras. Lalitalaya created mechanical men and arranged for the exhibition of a mock-duel between them; be created an artificial cloud and brought down heavy showers; with yantras, he exhibited magic; he devised a machine for war from which shafts as stout as pestles were discharged by him on the heads of elephants.

दृष्टेऽपि तस्मिन् विस्मयस्पृशो जनस्य " त्र्यं किल यन्त्रपुरुषैः द्वन्द्वैः युद्धमहीनमादर्शितवान्! त्र्रनेन किल प्रलिकजलधरधाराजलजालदन्नुरितमन्तारिक्षं कृतम्! एष किल यन्त्रमयमिन्द्रजालकं कृतवान्! एष किल संख्येषु प्रसंख्यानां युगपदेव मिनत्ति शत्रुहस्तिनां मस्तकस्थलानि मुसलमात्राभिरिषुमिः! प्रमुना किल द्रमिडभाषया शूद्रकचरितमुपनिबद्धम्! प्रस्य किल पित्रा यवनानप्यतिशयानेन क्षुधितोऽयमिति यन्त्रेरागभिधावितम्! प्रयं ततौऽप्यधिकः किल" इत्येवमासन् विकसितकुतूहलाः प्रलापाः!|

Lalitalaya is said to be master of all kinds of yantras; the varieties mentioned in this connection are six, *Sthita, Cara, Dhara, Dvipa, Jvara* and *Vyamisra*. Other texts speak of the classes of yantras as two and five and give the classifications somewhat differently. The *Sthita* or *Sthita* and *Cara*—stationary and mobile—is a classification going back to Kautilya. *Dhara* is plainly water-works; the manuscript gives the next as *Dvipa*, which had been wrongly construed also by some; *Dvipa,* I think, is an error for *Dvipa,* meaning elephant, and refers either to animal-shaped yantras or to special machines, such as we have already noted, employed against elephants in battle; it , means heat, may refer to machines involving the employment of fire. *Vyamisra is* a yantra,; par taking of the character of all these.

कल्पवृक्षक्रियाविस्मापितदुर्जयस्य मान्धातृनाम्नः स्थ्यपतेः प्रशस्तवास्तुशास्त्रार्थसारसामस्तय-संहारोन्मीलितप्रयोगतन्त्र ... वास्तुविस्तारकुशलः घरारावतिप्रासादविधिविशारदो यानासनशयनादिविकल्पनापटुः स्थित-चर-धार (रा)- द्वीप (द्विप)-ज्वर-व्यामिश्रसंज्ञानां षड्विधानाम् यन्त्राजां प्रद्वितीयप्रयोक्ता षट्-त्रिंशदाचार्यगुरौः प्रलड्कृतः ललितालयनामा समस्तसूत्रग्रही वर्धकी तज्ञकपज्ञप्रतीज्यः ज्ञत्रियैश्चच संस्कृतः (सत्कृतः) स्थपतिरभ्येत्य विरचिताञ्जलिरादृष्टिर्निर्दिष्टायां भूमावुपाविशत्!

प्रबसिता एव सर्वे नित्यप्रमादशैथिल्याभ्यां शिल्पातिशयाः, यतो प्रह त्वेवं प्रयोगलेशोऽपि विस्मयाय लोकस्य ! युष्माध्शां तु ब्रह्मेन्द्रपराशरप्रभृतिप्रगीतशास्त्रहृदयवेदिनां कियदिवैतन्नैपुराम्!

The machines referred to here are confirmed on either side by Kautilya in. his *Arthasastra* and Bhoja in the *Samaranganasutradhdra*. The, mechanical fighters are included in Bhoja; the artificial rain occurs in Kautilya in the, yantra called Parjanyaka, and the machine intended to smite the elephants' heads may be related to the Hasti-yantra advocated by Kautilya for use against an elephant corps. From a passage that follows it would appear that these yantras are dealt with in treatises associated with the authors Brahma, Indra- and Parasara, and that their vogue had become so reduced by long neglect that even-humble efforts in the line excited people's wonder.23

We now proceed to cite some texts on the existence and popularity, of a different category of yantras, till now we have been dealing with yantras having mainly the background of war. The yantras we shall now deal with are accessories of pleasure and entertainment, and more properly come, under household fittings and architectural engineering. Some of them are for thee reduction of human labor, some for sport and merriment,—toys- and gadgets of miscellaneous kinds for entertainment.24

We-may begin with Somadeva Suri, an encyclopaedic Jain writer, and his long religious poem, the *Yasastilaka Campu* written in South India in 949 A.D. In the first part of the work, Somadeva describes the hero resorting to the cool *yantra-dhara-griha* to spend the hot hours of the summer days. This park, fitted with mechanical fountains; is appropriately called by the commentator - *Kritrima-megha-mandira*, the artificial cloud pavilion. It is erected in the dense garden in an area provided with many canals. There is the stream for water' sports in the midst of which is a sandbank raised like a pavilion, provided with» a water-bed, *Salila-tulika*; nearby are numerous vessels containing fragrant water; at one end here is an *yantra-jala-dhara,* contrivance producing an artificial Waterfall; the water is taken through and thrown out of the mouths of figures of elephants, tigers, lions, snakes, etc.Other artificial works here are figures of celestial trees, *Kalpavrikshas*, with celestial damsels seated on them along with their lovers and figures of :cloud-damsels (*payodhara-purandri* or *Meghaputtalika)* giving shower- baths from their bosoms, figures of monkeys spouting water, statuettes of water- damsels, *(jaladevatas);* there are wind-damsels (*pavana- Kanyakas)*, wafting breezes with fly-whisks; and figures of ladies, scattering cool sandal-water all around. Somadeva Suri says of such a figure that if her hands were touched, shi would emit sprays through her nails; if her face, through the eyes and so on, a description which, as we shall see presently from its corroboration in "every detail in Bhoja's treatise, pertained clearly to fact and hot to mere imaginative fiction. That mechanical fountains were constructed as a necessary adjunct to all palaces is seen even in the casual descriptions in the dramas, the *Malavikagnimitra* (II. 12) and the *Nagananda* (III. 7) for example,

describing *jala- yantras*. The mechanical breeze-lady in the *yantradhara- griha* in the park has •her .companion within the bedchamber where

Somadeva Sari describes how near -the bed was a *yantra-putrika* plying a fan for the King's relief.

Many others of this class we shall be meeting in Bhoja.

Bypassing chronology a little, I shall now. Take up three available Sanskrit versions of- the *Brihatkatha.* The story thesaurus called *Brihatkatha* is on a par with the two epics, and can very well be called the great epic of popular fife. The version of Budhasvamin is the earliest, belonging probably to the 9th – 10th centuries, and coming from and based on a non-Kashmirian area and source. Budhasvamin has important information on aerial vehicles. Akasa-yantras, as he expressly calls them; it is also to be noted that he explicitly mentions the Yavanas as the knower of these Akasa-yantras, and the geographical area where he locates the architect and his exploits is also the part of the country where the foreign tribes had settled. The names of the architects too bear a strange complexion, suggesting their foreign origin.

The context where the description of the "yantra " occurs is the longing of the pregnant Vasavadatta; in the Kashmirian version of Somadeva there is only a line saying that her *dohada* (longing) was fulfilled by manifold contrivances, yantras, etc., arranged by the Minister Yaugandharayana. But

in Budhasvamin's version the context contains an elaborate digression devoted to the yantras. Vasavadatta yearned to see the whole world from above in an aerial vehicle (Sloka 190); Rumanvan, the commander-in-chief, at. once ordered .carpenters to manufacture a flying yantra (sloka 196). The carpenters say that they know only four kinds of yantras, made respectively with water, stone, mud and twigs; that it is the Yavanas who know the *Akasa-yantras,* and that they, for their part, have not even laid eyes on them.

Thereupon a Brahman told Rumanvan a story to illustrate how in the matter of the aerial vehicle, architects made a secret of their lore and uttered the falsehood that they knew it not.

With Mahasena, King of Ujjain and father of Vasavadatta, was an architect named Pukkasaka who once went out to Saurashtra along with the King's camp. There he came across a young architect, Visvila by name, who was verily' the equal of Visvakarman. To visvila's father, called Maya, Pukkasaka proposal that he desired to give his daughter Ratnavali in marriage to his son. The proposal wars agreed to and Pukkasaka was waiting for the arrival of his son-in- law. Once, after attending to his work, Pukkasaka returned rather late and to his surprise found none in his house eager to attend to his bath and his dinner.

On enquiry, he heard from his wife that a visitor had upset their home.; the. visitor had come over with some rice and asked that it be cooked for him ; they in the house had been burning toads of fuel and yet the rice would not even -moisten. Pukkasaka now understood that his son-in-law had arrived and desired to see the youth. Visvila issued out of the workshop and when the puzzled Pukkasaka asked him what that so-called rice was, Visvila revealed that they were fake, fine rice-like chisellings from the white wood of the Karaghata tree. The marriage of Visvila and Ratnavali was then celebrated. After a time, Visvila learnt from his brothers-in-law their anxiety about feeding in their house one more family member in the form of the son-in-law; Visvila at once repaired to the forest, cut down certain kinds of wood and manufactured out of them yavana-machines, as also manifold household utensils conducive to health and longevity, according to the principles laid down in *Vrikshayurveda* (225); he sold these for thousands of pieces of money and presented the gains to his father-in-law.Once Pukkasaka sadly told his family of his impending departure for Benaras whither King Mahasena had ordered him to go to build a temple for his friend King Brahmadatta of Benaras. Visvila asked leave to deputize for his father-in-law and, with the King's permission and accompanied by a retinue, he departed for Benaras. At the end of every day's journey, however, Visvila would vanish somewhere and slip back inconspicuously into the camp. At his home, Ratnavali shortly became pregnant, to the surprise and agony of her parents. That news reached the King who, putting two and two together, explained that every night, Visvila, according to the report of the men of his retinue, would mount a machine-cock, *Yantra-kukkuta*, fly away somewhere and, stealthily and with shrouded face slip back into his bed in the small hours of the morning. Once he had been forced to return late in the morning and, confessing to his friends about his nocturnal visits to his wife by an aerial vehicle, he had begged them not to inform any, architects or laymen, of the secret of his aerial vehicle, which in fact could not be understood by non-Yavanas; if that knowledge became public, the Akasa-yantra would become a cheap affair like a cot.

It may be mentioned here .that the association of skilled works, with Yavanas had become so well known that we find it mentioned in Tamil literature also. In the fragmentary Tamil version of the Brihatkatha, by Konguvel, Yavana carpenters are included among the workmen of many types responsible for a chariot of high workmanship used by Udayana (I. 58.40}; and in the Buddhistic epic Manimekhalai, in an almost similar list of workmen mentioned as responsible for a garden house of great architectural charm, Yavana carpenters are again found That explained how Ratnavali came to bear a child. Soon the temple at Benaras was finished and Visvila returned.

King Mahasena now pressed Pukkasaka for knowledge of the "'Akasa- - yantra." Pukkasaka replied that he had not taught Visvila; he, in; fact, did not know it, the Yavanas being the custodians of that knowledge ;

The King would not believe his words and pressed him, whereupon Pukkasaka pressed Visvila. Visvila pretended to reveal the secret, but that night he woke up-his wife and gave her an ultimatum; by pressing him for knowledge of the flying-machine, her father was virtually driving him to his own home, that she had to choose between father and husband, and that, so far as he was concerned, he would give her up rather than the secret. She took little time to decide and in a moment they were off on the machine-cock:

यानं कुक्कुटसंस्थानमास्थाय सह भार्यया !
रात्रावाकाशमुत्पत्य स्वस्थानं विश्विलो ययो !!

Note- Refer machine no 56 Bhadrashwa Yantra . This Machine Cock is similar to it in function

Having told the story, the Brahman told the commander-in-chief Ruman- van that architects made such secrets of their knowledge, that all of them might well be bound and beaten till they agreed to make the aerial car. As Rumanian was putting that advice into action, there came a fresh architect who *offered* to manufacture an aerial car.
The new *Shilpin(Engineer)* asked Rumanvan to collect the materials. When these were assembled and the work was to start, the old *Shilpins*

suggested to the new one that he ascertain from Rumanvan the seating capacity that was required. The mention of this deserves to be noted, as also the further observation that there had been cases in the past in which aerial cars had been made without regard to seating capacity and they had come to ruin, with the result that their makers had been cruelly dealt with by the Kings:—The new *Shilpin* replied that the yantra that he was going to make was of a superior type quite different from the productions of the stupid architects to whom they referred, and that the seating capacity of his vehicle was not limited. He made accordingly an aerial car of the shape of Garuda. Vasavadatta and Udayana mounted it with their retinue, roamed about, called on Padmavati's brother in Magadha on the ea«t and on Vasavadattas own parents at Ujjain on the west and returned to Kausambi.

This fulfillment of the Queen's *dohada* during her pregnancy, for an aerial flight, has an echo in Jain *Kavya* literature also; in Vadibhasimha's *Gady cintamani* and *Kshatrac&gamani*[81], carpenters make a peacock-like aerial car (*Mayura-yantra*) for the pleasure flight of Queen Vijaya.

On smaller mechanical objects also we have some information in. the *Slokasamgraha* of Budhasvamin (Chapter XIX, p. 287). At Campa, when Naravahanadatta was staying with Gandharvadatta, they heard the story of the origin of a local water festival. An old King of Campa had a Queen who's' *dohada* during pregnancy expressed itself as a desire to move about in waters' filled with all sorts of aquatic beings. For her sake the King dammed a river, widened it into a big lake and fitted it with wooden replicas (*Daru-yantras*) of' crocodiles, fishes, etc., which moved freely in the water, and there let her sport in a vessel shaped like an aerial vehicle.

तत्र नन्कादिसंस्थानदारुयन्त्रनियन्त्रिते!

विमानाकारपोतस्थौ तौ राजानौ विचेरतु:!!

Of the two Kashmirian, versions of the *Brihatkatha,* Kshemendra's is very brief, and Somadeva's long enough for us to glean much information about- yantras. These two works may be noticed together. We have here material bearing on three classes of yantras, dolls and entertainment pieces, mechanical men and women, and aerial vehicles.

In the course of the Madanamanchuka story in VI. 3, Somadeva narrates in his *Kathasarisagara* the episode of Somaprabha, the daughter of Maya, who takes a fancy to Madanamancuka and becomes her great friend; Somaprabha and Svayamprabha, of *Ramayana* and *Pailcapsaras* fame, was two daughters of the Asura architect Maya. One morning Somaprabha calls on Madanamancuka with a wonder-box full of various kinds of mechanical dolls. Maya himself had taught Somaprabha how to make these yantras, mechanical wooden toys;

Then follows a description of four of these toy-yantras: By striking, at a, pin, one makes a yantra jump up in the air and it comes down with a garland; another similarly comes back with a cup of water; a third dances; and the fourth sits up and gossips. After entertaining Madanamancuka with these; Somaprabha leaves the box in the former's care and departs. Next day, when she calls again, Madanamancuka introduces her to her own parents. The two: then go to the royal park where one of the toys brings forth a Buddha image and materials for worship. Hearing of this, the parents rush to see the wonder and on the King asking her about the yantras and how they go into action, Somaprabha. Gives a short account of the yantras that her father had devised; the treatment here has some parallels with that in Bhoja's work.

Somaprabha. Says that, just as the universe is made up of five elements, yantras are also based on the five elements of earth, water, fire, air and ether. The yantra based on earth-materials *(prithvi-pradhana)* undertakes activities like shutting doors; a *water-based yantra* will be as lively as a living] organism; a *fire-yantra* emits flames; an *air-yantra* moves to and fro; and the element of ether serves to convey the sound generated by these yantras.

Somaprabha adds that there is a *super-yantra* called *Cakra-yantra* with miraculous powers which her father did not teach her, but in the description of this there is an obvious mix-up of mythology. Then, with the parents' permission, Somaprabha takes Madanamancuka in an aerial car to her own father's place, where her sister Svayamprabha is living, and returns to Takshasila. Kshemendra's account of this episode is brief, comprising about a dozen verses in his *Brihatkathamanjari* (VII. 195-207). The other story in which yantras figure mainly is in the *Ratnaprabhd-lambaka,* VII. 9, the story of Karpurika in the city of Karpurasambhava, of whom Naravahanadatta has a dream. The Prince starts out in search of her, in the company of Gomukha. *En route*, they come to a Hemapura, where, as Naravahanadatta is going along the bazaar street, he comes across everything pertaining to a city, shops, things, servants, men and women, but their speechless movements and activities reveal to him the wondrous fact of their all being robots (10, 11). He then makes his way to the palace where he finds the only sentient being sitting on a throne like a king, and, like the soul presiding over the body and senses, manipulating the mechanical city and being served on all sides by his mechanical servants. On being asked by Naravahanadatta, the mystery man of this machine city recounts his story

He, Rajyadhara by name, and his elder brother Pranadhara, were originally residents of Kanchipura where King Bahubala (Mahabala in Kshemendra) was ruling. Both of them were architects and adepts in the manufacture of magic yantras made of wood, etc., and devised originally by Maya. It may be noted in passing that, according to Dandin, Kanchi had some architects who were adepts in the making of yantras. The elder brother sought the company of courtesans and squandered his as well as his younger brother's property. Reaching the end of his material resources the elder brother thought of theft and harnessed his skill in yantras for that purpose. He devised a pair of wooden swans which could move along a rope-contrivance manipulated from one end. The other end of the rope was tied to the window of the King's treasury into which, night after night, these swans were- sent; their beaks, which were put into action, removed the lids of jewel boxes and picked up 9eift« jewels and then the swans were again moved back to their original place. This mysterious theft; was going on, and the King ordered an all-night vigil to catch the culprits. The swans were seen doing this dexterous job and were detached from the rope, which suddenly sagged; the fastening nail became loose and Pranadhara understood at once that the theft had been found out. He asked his brother to accompany him to a far-off place to escape being caught as thieves early in- the morning. Pranadhara said that he bad with him an aerial vehicle that could in a single sweep rocket across 800 *yojanas* and got into it immediately, with his family. As the seating capacity of that yantra bad been reach ed by the crowd that entered it, Rajyadhara left in his vehicle, a *vata-yantra- vimana,* as it is here styled, which he himself had made; that machine took him at one jerk over 200 *yojanas,* and with a second propelling, over another 200 *yojanas.*

That brought the younger brother, the narrator, to the city of Hemapura. When he reached it, it was an abandoned place; he thought of peopling it with, the help of his mechanical skin and created a yantra-population. Next day, he learnt from Naravahanadatta the search that the latter was on, and helped him with an aircraft which took him to the city of Karpurasambhava There the Prince found the heroine of his dream, Karpurika, and married her. When he desired to return home with his new bride his father-in-law revealed that he too had a visitor-architect in his city who could provide him with an aircraft.. It happened that the visitor-architect was no other than Pranadhara, the elder brother of Kanchi. The vehicle that Pranadhara gave him was a veritable flying fortress, as it could lightly bear a thousand passengers (328), and in that they returned home, touching Rajyadhara's city *en route.*In Kshemendra's brief narration of this story (XIV. 459-508), the thieving swans are mentioned as many, the elder's aircraft is called a *Yantracakra,* and the younger's is said to possess double the speed of the elder's.

There are also two other contexts in the *Kathasaritsagara* where the aerial yantra figures (the story of Pushkaraksha and Vinayavati and the story of Somaprabha and the three suitors).

This gives brief idea of multipassenger aerial vehicles of South Indian Kings of Kanchi & Mahabalipuram in 7th century AD

Chapter IX

Missile Technology of Bhrigus

Bhrigus belongs to the one of the most important creator family of Aryans . There are such seven creators . Bhrigus are very ancient & most powerful amongst them . Bhrigu literally means planet Venus . Many great sages in this lineage had their names after planet venus . There were also called Kavis , as it was the name for venus in sumerian language .Bhrigus were brahmin-warrior emperors in Babylon-Assyria or modern Iraq-Iran region . For this reason , in Aryan Puranas , Bhrigus are mentioned as Gurus or ministers of Asuras, people who living in Assyria .They were exceeded in wisdom & warfare both . These two Ws gave them upper hand over most of the brahmins & kshatriyas during that time . They were in continuous war with Sythians Kshatriyas who were ruling present day Pakistan to Middle India including present Gujrat, Rajasthan & Maharashtra . Haiheya were mighty race of sythian kshatriyas . Because of this continuous war & for survival Bhrigus had to develop new technologies in weapons & warfare . Missile technology is one of such great inventions of Bhrigus .

In lineage of Bhrigu , many such warrior sages took births , Shukra or Ushana was one of them , who wrote great works like Aushanasa Dhanurveda . He was first who formed Displined Army with hiearchy . He called it as Sanjeevani Vidya . As per this military science , General, army cheif, solider, fighter were posts , replacable by another trained person . This was totally against the principles of Deva army in which all titles & posts were as per births . If General or Senapati dies in a war, none was there to replace him & all Devas will run away . Many times Asuras won over by Devas by just capturing their King Indra . As there was no system to replace , rest Devas will run in fear & accept defeat . Many years Deva army which were on eastern side of Sindhu river did not figured this out . Finally they sent their spy Kacha to Shukra Guru to learn this Sanjeevani Vidya by trickary . After getting it Indra changed his army & made Kartikeya and his Yaudheya Ganas as Generals . Sanjeevani Vidya was path breaking contribution in war science during that time . And here we are talking of time min 3000 BC

After few centuries in the lineage of Shukracharya was born great Urva ,
proud Bhargava who took revenge of death of his mother caused by
Kshatriyas . Urva was first known Bhrigu who developed missiles . For his
invention , Missile fuel was named after him as Urvagni . Rama of
Ayodhya used his Shataghni missile run on Urvagni to attack city of Lanka
in his battle against Ravanana . Ramayana of Valmiki mentions it .
Shataghni is another name for missile or rocket .

'और्वाग्निं प्रोथितं कृत्वा शतघ्नीं गुडकैयुं ताम् ।
चिक्षिपुर्भुं जवेगेन लङ्का मध्ये महाखनः' इति ।

After Urva , great Chyavana was born . He took glory of Bhrigus in
greater heights especially in metallurgy . His son was Jamadagni . His
most life was spent with fierce battle with Haiheya king , Kartaveerya
Sahasrarjuna . Kartaveerya who built great elephanta caves on island near
Mumbai , was mighty king who even defeated Ravana of Lanka . He ruled
over Maharashtra, Gujrat , Rajasthana & present day Pakistan upto Iran .
Jamadagni was killed by him in personal battle . This was devastating
attack on Bhrigus . But Bhrigus had warrior blood in them along with
wisdom . His son Parashuram took revenge of death of his father . For this
he received in a boon new war science , Jamadagnya Dhanurveda ,
directly from Shiva . By using weapons of this new technology , he killed
Haiheya King Kartaveerya & destroyed almost all Sythian Kshatriya races
in western & northern India . Because of advanced weapons Parashuram
possessed , for many years no Kshatriya from India was able to challenge
him , even they gave their wives to Bhrigus to recieve progeny . This one
side battle went for many centuries . Bhrigus did not taught this knowledge
to Kshatriyas till then . First Kshatriya to receive this knowledge was Kuru
King , Devavrata Bhishma of Hastinapura . Parashuram imparted him all
his weaponary .

In Jamadagnya Dhanurveda , astra is defined as a weapon which is thrown either by secret mechanism like yantra or by explosion by fire .

अस्यते क्षिप्यते यत्तु मन्त्रयन्त्राग्निभिश्च तत् ।
अस्त्रं तदन्यतः शस्त्रमसिकुन्तादिकं च यत् ॥ १९१ ॥
अस्त्रं तु द्विविधं ज्ञेय नालिकं मान्त्रिकं तथा ।

This astra is classfied into two types
1. Hollowed weapons like guns , revolvers, rocket launcher guns
2. Thrown by mechanism like explosions of fire , gases & water

ब्राह्मं नारायणं शैवं ऐंद्रं वायव्यवारुणे ॥
आग्नेयं चान्यदस्त्राणि गुरुदत्तानि योजयेत् ॥ जामदग्न्य

Agneya Astra , the missile launched by fire was trademark Bhrigu Astra . Even now , India named its missle program as Kshepanastra & Agni is India's most advanced missile which is nuclear warhead carrier .

Nalikastra is said to be also of two types

नालिकं द्विविधं प्रोक्तं बृहत्क्षुद्रप्रभेदतः ॥
लघुदीर्घाकारधाराभेदैर्बहुविधं भवेत् ॥ जामदग्न्य.

Short size guns & Big grenade launcher

नालाग्निचूर्णसंयोगाल्लदये गोलनिपातनम ॥ ३१६ ॥
नालिकास्त्रेण तद्युद्धं महाह्लासकरं रिपोः ।

Targetting the enemy by grenade attack by help of launcher weapon is said
to be the most destructive type of warfare .
In both launcher guns & missile weapons , two parts are most important .
Rocket fuel & Projectile .
For Nalikastra , lead bullets , iron nails , explosives are used .

नालास्त्रं शोधयेदादौ दद्यात्त्राग्निचूर्णकम् ॥ २०९ ॥
निवेशयेत्तदण्डेन नालमूले यथा दृढम् ।
ततः सुगोलक दद्यात् ततः कर्णेऽग्निचूर्णकम् ॥ २१० ॥
कर्णचूर्णाग्निदानेन गोलं लक्ष्ये निपातयेत् ।

For rocket fuel different kinds of combinations of Salt Petre & sulfur were
used

सुवर्चिलवणं पंचपलानि गंधकं पलं ॥
अंतर्धूमविपक्काकंसुह्वाद्यंगारकं पलं ॥
शुद्धं संग्राह्य संचूर्ण्य समील्य प्रपुटद्रसैः ॥
शुद्धार्काणां रसे तस्य शोषयेदातपादिना ॥
स्पृश्वाश्करंरविचितजामदग्न्यमिति स्मृतं ॥
नालास्त्रं चाग्निचूर्णं च गंधांगारश्च कथ्यते ॥ जामदग्न्य.

105

Some also carried Mercury Sulfides , liquid mercury , lead . Some also used Pranaagni ie Oxyegen as rocket fuel .

अंगारस्यैव गंधस्य सुवर्चि लवणस्य च ॥

शिलायाहरितालस्य तथा सीसमलस्य च ॥

हिंगुलस्य तथा कांतलोहस्य कर्पूरस्य च ॥

जंतोर्निल्याश्व सरलनिर्यासस्य तथैव च ॥

समन्यूनाधिकैरंशैराभिचूर्णान्यनेकशः ॥

कल्पयंति च तद्विद्या चंद्रिकास्कोटकादिषु ॥

This technology from Jamadagnya Dhanurveda went to NAZI Germany & resulted in invention of V1 & V2 rockets . In Introduction of this book, we have seen how Aryan sciences in sanskrit had reached to Europe in early 20th century . German advances in fighter planes , guns , tanks and rockets was direct result of Aryan technology taken secretly from India .

V2 & V1 rockets

V1 Rocket which had pulse jet engine on it

This technology transfer gave NAZI germany decisive advantage . London was burned to ashes as these rockets were falling from the sky . US & Russia took this technology after WWII ended , used it to start their Space Program . This is not the subject about space programs , but NAZI germany started its own Space Program in 1930s . V1 & V2 were part of Vril series of rockets only but only for open wars . Other rockets of Vril series were under secret space program .

Even Wikipedia mentions about V2 rocket going to outer space . MW 18014 , V2 rocket was first man made rocket to enter into outer space at height of 176 km . This happened in June 1944 as per Wiki .
Clearly real space program was hidden from masses .

US took this technology and launched its Jet Propulsion Laboratory for Space research & Space Rocket launch . Now many nations have their space programs , but all together could not surpass the wisdom of NAZIs and Bhrigus . Missile Technology of Bhrigus was such a game changer for modern world . Jamadagnya Dhanurveda hails missile weapons as best weapons for winning the war .

मन्त्रेरितमहाशक्तिबाणाद्यैः   शत्रुनाशनम् ॥ ३३५ ॥
मान्त्रिकाख्येन तद्युद्धं सर्वयुद्धोत्तमं स्मृतम् ।

References & Blibiography :

Prachin Vimana Kalecha Shodh by Shivkar Bapuji Talapade 1908
Guru Mantra Mahima (Gujrati) by Shivkara Bapuji Talpade 1915
Prachin Hindi Shilpashastra by K V Vaze
Brihad Vimana Shastra of Maharshi Bharadwaja published by
Sarvadeshika Arya Pratinidhi Sabha 1956
Prasthana Trayee of Madhusudana Saraswati published by V B Soobiah &
Sons 1931 under guidance of Shri Subbaraya Shastri
Bhautika Kala Nidhi monthly Jan 1911 issue
Black Sun : Aryan Cults, Esoteric NAZIsm and The politics of identity by
Nicholas Goodrick Clarke
SS Brotherhood of the Bell : The Nazi's incredible secret technology by
Joseph P Farrell
Das Vril Projekt by Jurgen Ratthofer / Ralf Ettl
The truth about Wunderwaffe by Igor Witkoswki
Shilpa Sansara Monthly , Year 1 Vol 17 April 1955
Manuscripts of Subbaraya Shastri in possesion of G Venkatachala Sarma
Jamadagnya Dhanurveda
Shukraniti
Aushanasa Dhanurveda Sankalanam by Arsha Granthavali Lahore 1923
Samarangana Sutradhara
Avanti Sundari by Dandin
Brihat Katha Sara
Yantrarnava
Agastya Samhita